HVAC Procedures and Forms Manual

Second Edition

HVAC Procedures and Forms Manual

Second Edition

by Herb Wendes, P.E.

River Publishers

Routledge
Taylor & Francis Group
LONDON AND NEW YORK

Published 2020 by River Publishers
River Publishers
Alsbjergvej 10, 9260 Gistrup, Denmark
www.riverpublishers.com

Distributed exclusively by Routledge
4 Park Square, Milton Park, Abingdon, Oxon OX14 4RN
605 Third Avenue, New York, NY 10017, USA

Library of Congress Cataloging-in-Publication Data

Wendes, Herbert.
HVAC procedures and forms manual/by Herb Wendes.--2nd ed.
p. cm.
Includes index.
ISBN 978-8-7702-2224-2 (electronic)
1. Heating--Estimates--Forms. 2. Ventilation--Estimates--Forms. 3. Air conditioning--Estimates--Forms. 4. Buildings--Mechanical equipment--Installation--Estimates. 5. Buildings--Energy conservation.

TH7335 .W4597 2001
697'0029'9--dc21

2001040687

HVAC procedures and forms manual/Herb Wendes.
First published by Fairmont Press in 2002.

Routledge is an imprint of the Taylor & Francis Group, an informa business

0-88173-387-3 (The Fairmont Press, Inc.)
978-0-8247-0923-5 (print)
978-8-7702-2224-2 (online)
978-1-0031-5204-0 (ebook master)

Contents

INTRODUCTION ix

CHAPTER 1

HVAC Testing and Balancing Procedures and Forms i

Air Balance Test Reports 1

General Procedure for Balancing 3

Equipment Test Reports 11

Air Balancing Procedure 12

Testing and Balancing Report Cover 13

Fan Test Report 16

Outlet Air Balance Report 18

HVAC Systems Recap Sheet 20

Pitot Tube Traverse, Rectangular Duct 22

Pitot Tube Traverse, Small Round Ducts 24

Pitot Tube Traverse, Large Round Ducts 26

Hydronic Balancing Procedures and Reports 28

Hydronic Balancing Procedure 28

Pump Test Report 32

Water Balance Report, Pressure Differential 34

Water Balance Report, Temperature Differential 36

Chiller Test Reports 38

CHAPTER 2

HVAC Energy Auditing Procedures and Forms 41

Sample Commercial Audit Procedures and Forms 41

Energy Audit Procedures 41

In Depth Energy Audit Procedure 42

Readings Required for Thorough HVAC Audit of System 46

Average Annual Energy Performance in Btu's Per Square Feet 48

Sample Audit and Forms 49

Drawing of Suburban Office Building 50

HVAC Energy Auditing Report Cover 51

Building and System Description 52

Electrical Consumption History 56

Equipment Schedules 58

Electrical Consumption Graphs 60

Equipment Test Reports 62

Electrical Loads Per System 66

Electrical Loads Project Recap 68

Heat Loss Calculation, Seasonal 70

Cooling Load Calculation, Seasonal 72

Peak Heating, Cooling and CFM Per Area 74
Energy Cost and Btu Profiles 76
Residential Audits 78
Energy Consumption History 78
Outside Wall Energy Evaluation 80
Ceiling and Roof Energy Evaluation 82
Windows and Doors 83
Cold Floor Energy Evaluation 84

CHAPTER 3
Indoor Air Quality Procedures and Forms 85
Problems 85
Specific Causes of Indoor Pollution 86
Testing, Setting and Controlling Outside Air 88
Carbon Dioxide Levels 89
Indoor Air Quality Instruments and Testing 90
Humidity 90
Particulate Contaminants 91
Formulas 93
Indoor Air Quality Report 95
Building Ventilation and Pressure Report 96

CHAPTER 4
Engineering Calculation Procedures and Forms 97
Heat Loss Calculation Form 98
Cool Load Calculation Form 100
Peak Heating/Cooling/CFM Per Area 102

CHAPTER 5
HVAC Cost Estimating Procedures and Forms 105
Introduction 105
Sheet Metal Estimating 106
Requirements of a Good Sheet Metal Estimator 106
Efficient Estimating Procedures 112
Estimating Procedure Diagram 115
Scope of Complete Sheet Metal Estimate 116
General Estimating Forms for Both Sheet Metal and Piping 117
Estimate Coversheet 117
Job Description and Budget Costs 118
Bid Checkoff Sheet 120
Quotation Call List and Price Comparison 124
Quantity Takeoff and Extension Sheet 126
Estimate Summary Sheet 128
Telephone Quotation 130
Bid Recap Sheet 132
Bidding Record 134

Calculating Labor Costs Per Hour 136
Company Markup Calculation Sheet 138
Duct Takeoff and Extension Sheets 140
Per Piece Duct Takeoff Sheet 142
Per Pound Duct Takeoff Sheet 143
Piping Estimating 144
Requirements of a Good Piping Estimator 144
Piping Takeoff Sheet 150
Computerized Estimating 152

CHAPTER 6
Hitting Home Runs in Technical Management
Introduction 157
Faced with A Multitude of Problems 158
Understand What Constitutes A Problem 162
Multiplicity Thinking; Avoid Partial Views 164
Be Acutely Aware of the Principles and Procedures 164
Outline of the Principles and Procedures 165
Vibrations in High School Building, Sample Case History 168
Become thoroughly Familiar with the Problem 173
Examine Possible Causes and Suspend Judgment until Verified 176
Establish and Evaluate Objectives 178
Generate Multiple Options and Hold Back on Final Choice 180
Information and Planning Worksheet 182
Evaluate Options, Compare against Objectives, Make Decision 183
Write it Down 184

APPENDIX A—Blank Forms 187
A. Full Size Blank Testing and Balancing Forms for Your Use
Testing and Balancing Report Cover 188
General Information Form 189
Fan Test Report 190
Outlet Air Balance Report 191
Air Systems Recap 192
Pitot Tube Traverse, Rectangular Duct Form 193
Pitot Tube Traverse, Small Round Ducts Form 194
Pitot Tube Traverse, Large Round Ducts Form 195
Pump Test Report 196
Water Balance Reports 197
Chiller Test Report 199
B. Energy Audit Forms for Your Use 200
Coversheet for Energy Auditing Report 201
Building and System Description 202
Electrical Consumption History Per Month 203
Fan Test Report 204
Pump Test Report 205

Chiller Test Report ... 206
Electrical Loads per System ... 207
Electrical Loads Project Recap ... 208
Heat Loss Calculation ... 209
Cooling Load Calculation ... 210
Peak Heating, Cooling and CFM Per Area ... 211
Energy Consumption History ... 212
Outside Wall Energy Evaluation ... 213
Ceiling and Roof Energy Evaluation ... 214
Windows and Doors ... 215
Cold Floor Energy Evaluation ... 216
C. **Indoor Air Quality Forms** ... 217
Indoor Air Quality Report ... 217
Building Ventilation and Pressure Report ... 219
D. **Engineering Calculation Forms for Your Use** ... 220
Heat Loss Calculation ... 220
Cooling Load Calculation ... 221
Peak Heating, Cooling and CFM Per Area ... 222
E. **HVAC Cost Estimating Forms** ... 223
HVAC Estimate Report Coversheet ... 223
Job Description and Budget Costs ... 224
Bid Checkoff Sheet ... 226
Quotation Call List and Price Comparison ... 227
Quantity Takeoff and Extension Sheet ... 228
Summary Sheet ... 229
Telephone Quotation ... 230
Bid Recap and Markup Sheet ... 231
Bidding Record ... 232
Calculating Labor Costs Per Hour ... 233
Company Markup Calculation Sheet ... 234
Per Piece Duct Takeoff Sheet ... 235
Per Pound Duct Takeoff Sheet ... 236
Piping Takeoff Sheets ... 237
F. Hitting Home Runs in Technical Management Forms ... 238
Problem Work Sheet ... 238
Cause Work Sheet ... 239
Objectives Work Sheet ... 240
Options Work Sheet ... 241
Information and Planning Work Sheet ... 242

APPENDIX B—Charts and Formulas ... 243
Abbreviations and Symbols ... 244
Association Abbreviations ... 245
Air Flow and Pressure Formulas ... 246
Changing Fan CFMs and Drives ... 248
Air Heat Transfer Formulas ... 249

Changes in State of Air Formulas 250
Air Density Correction Factors 252
Converting Velocity Pressure Into Feet Per Minute (Standard Air) 253
Converting Velocity Pressure Into Feet Per Minute (Various Temperature) 254
Hydronic Formulas 255
Psychrometric Chart 256
Motor Amp Draws, Efficiencies, Power Factors, Starter Sizes 257
Conversion Factors 258
Fuel Heating Values 259
Steam Formulas 260
Electrical Formulas 261
Refrigeration Temperature/Pressure Chart 262
Properties Of Saturated Steam 263

INDEX 265

Introduction to Second Edition

Hitting Home Runs in Technical Management

This second edition of *HVAC Procedures and Forms* has been expanded to include a new section titled *Hitting Home Runs in Technical Management.*

> *The new chapter covers the management of problems, trouble shooting and decision making in the HVAC industry in a more effective, systematic, and thorough manner. Included are the principles and procedures of the home run process, an actual complete case history, and forms which complement the procedures and help guide you through the process.*

The five other sections of the manual provide a quick reference and guide on procedures, charts and forms in some of the most critical areas of operations in the HVAC industry, *testing, balancing, energy auditing, indoor air quality, load calculations, and cost estimating.*

The manual is written for contractors, building engineers, and technicians, as well as for balancers, maintenance personnel, and servicemen. It is a valuable guide for managers and supervisors for planning, scheduling, controlling, monitoring, and resolving problems quickly and effectively in mechanical systems.

- The comprehensive procedural sections guide the manager or technician effectively through testing, trouble shooting, auditing, air quality, and problem solving operations.

- The forms are concise, comprehensive and optimally organized for easy gathering of data and for reference. They prompt the user for essential information needed in the different areas of work. They serve as a check-off list to insure completeness and effectiveness of operations.

The procedures and forms in this manual were created by the author over a period of twenty-five years. He personally applied the procedures and the forms successfully in the field and office. He tested, modified and developed them and proved them to be meaningful and accurate. Thousands of engineers, technicians, balancers, maintenance personnel and contractors have applied the procedures, used the forms, and provided positive feedback. They have been used in schools and for training guides.

Because it would take many years to develop valid, workable, analytical procedures and forms, and because this information is now compiled in a single reference—the user will save countless hours in performing the technical operations, and in assimilating and organizing the data, on virtually all types of mechanical systems and buildings.

Chapter 1

HVAC Testing and Balancing Procedures and Forms

AIR BALANCE TEST REPORTS

Test reports are an absolute necessity in balancing. They prevent chaos, errors and generally a big mess of indecipherable and incomplete paperwork.

Test reports enable the balancer to keep things organized, clear and neat. They show the procedure and guide the sequence of work. They act as a constant reminder of what information is needed.

They are excellent records to refer back to. The balancer can go back and can refer to the reports to see how the systems were, how they should be, and what is actually different. He can quickly rebalance a system. Reports can be used as the basis of redesign and/or problem analysis.

Organized forms are easy to read, not only for the balancers, but for others such as the designers or building engineers. Most of all, forms ensure better, faster and more efficient balancing, and fewer call backs.

Types of Reports

There are a number of different types of test reports used in air balancing. The most widely used and critical ones are the *Fan Test Report* and the *Outlet Air Balance Report*. These provide the basis of essential data on the fan and the balance condition of the outlets.

The *Pitot Tube Traverse* sheet is more of a worksheet than a report that is distributed or kept for record purposes.

Larger, more complicated projects require *General Information Sheets* and *System Recaps* to tie everything together and keep track of the balancing progress.

Schematics of the systems, numbering the outlets, clearly showing the routing, etc. are frequently needed for larger systems—systems that are spread out, messy blueprints, and systems for which no blueprints are available.

Procedure for Preparing Reports

1. Gather the following information:
   - Blueprints
   - Specifications
   - Fan Submittal Drawings
   - Air Handling Unit Drawings
   - Grille Schedules
   - K Factor Data

2. Study the plans and specifications. Read the equipment schedules on the blueprints.

   See what types of systems and equipment there are, how the ductwork is routed, where dampers are located, what the size and capacity of outlets are.

   Read what the specification requirements are on balancing.

   - What degree of accuracy is called for, +/–5%, +/–10% or something else?
   - Do you have to balance in summer or winter?
   - Should heating and cooling be on or off?
   - Should the outside air damper be open or closed? What instruments and procedures are required?
   - Is certified balancing required? Can the contractor balance or must it be an independent balancing company?
   - Should the building be occupied or not?
   - Are new filters required?
   - Are temperature readings required?

3. Plan how the system can be best balanced and determine what instruments are to be used.

4. Fill out the general information sheet. Describe the types of systems with key descriptive terms such as low pressure, high pressure, variable air volume, single zone, roof top, split system, return plenum ceilings, etc.

Indicate the manufacturer of the grille, diffusers and terminal boxes. Describe the types required and furnished.

Checkoff the instruments to be used and indicate the models and last calibration dates.

Checkoff the status of completion of the building and systems.

GENERAL PROCEDURES FOR BALANCING

Low Pressure Constant Volume Systems

The most common type of HVAC system used in the United States is the low pressure constant volume supply system. There are hundreds of thousands of stores, supermarkets, restaurants, offices, factories and so on, which employ them. Possibly 60 or 70 million of the 88 million residences in the country are low pressure constant volume. Many of the systems in institutional buildings such as hospitals and schools are of this type.

There are three categories of low pressure constant volume systems, single zone, reheat and multi-zone. The most predominant of these is the single zone.

The low pressure constant volume system is defined as follows:

1. The total system *static pressure* doesn't exceed *2* or *2-1/2* inches water gauge.
2. Air *velocities* don't exceed between *2000* and *2400* fpm.
3. The systems are *constant air volume* and *variable temperature. To* control the temperatures in the spaces, the heating and cooling equipment is cycled on and off or modulated in order to vary air flow and space temperatures.

Basic Principles

1. Start at the *heart* of the system, which is at the fan, and make sure it is pumping correctly before balancing duct runs and outlets.
2. Make sure the system is clean before balancing, that all filters, coils, strainers, duct-runs dampers, louvers etc. are clean and unclogged.
3. Make sure the system is *open before* balancing, that all dampers behind grilles and supply diffusers, manual balancing dampers and fire dampers are open and that the control air dampers are set in the correct positions.
4. Make sure the air distribution system is properly *sealed* and that no duct end caps are left off, no duct runs are unfinished and no outlets are not connected. Make sure that connections and seams are sealed if required.

5. Balance on *maximum air flow* mode whenever possible and with *wet cooling coils* or assimilated conditions for actual operating maintenance.

6. *Proportionate balance.* Use this systematic method of balancing based on proven physical laws of fluid flow. It insures that the least amount of *resistance is* being input into the system to achieve the required air balance. It is a method which requires the least amount of balancing time and is the most efficient available. And the biggest feature of proportionate balancing is that the fan flow can be increased or decreased after a system outlet balance, without losing the balance, thus avoiding having to rebalance the outlets.

   The general procedure for testing and balancing low pressure constant air volume (CAV) systems is as follows:

Preliminaries

Study the plans, specifications and equipment drawings to become familiar with the systems. A determination must be made of the best method to balance the systems, and appropriate instruments must be selected and checked out.

Prepare test reports, study plans and specifications. The first stage in the testing and balancing procedure is the preparation of test reports. Equipment test report sheets must be completed for each system. Outlets must be listed on air balance sheets in the sequence of balancing together with their types, sizes, Ak factors if required, design air quantities and velocities.

Check that building and systems are complete and operational. After the reports are prepared, inspect the job site to see that the building and systems are architecturally, mechanically and electrically ready to be balanced and they are complete and functional.

Invariably, new buildings may be only half ready when balancing starts, and in fact, it is the balancer's quality control check that uncovers a multitude of missing or incorrect items. As the balancing technician inspects each system he must report the inadequacies, see that corrective action is taken and move onto the systems that are ready for balancing.

After you have determined which systems are truly complete choose the first one to balance and proceed with an in-depth equipment check out.

Check Heart of System First

Start the actual testing and balancing *process* at the *heart* of HVAC systems, at the *fan.*

If the heart isn't working right, the rest of the body can't perform as it should. Just as a doctor checks your blood pressure and pulse rate, you must check the fan's pressure and rpm rate.

The *motor* on the fan is the organ that *drives* the fan and its electrical characteristics must be checked out and it must be protected. Hence, the first phase in the testing and balancing process is to check five items at fan:

1. Motor amp draw and thermal overloads
2. Fan rpm
3. Fan suction and discharge pressures
4. Pressure drops across components
5. Total air flow at fan

Then after the heart of the system is checked, adjusted and running properly—and only then—should the *outlets* and *duct-runs* be read with instruments and balanced.

Check Motor and Starter

1. *Motor Nameplate.* Since the weakest link in the system is the motor, it is imperative that it be protected. Check the motor nameplate first for maximum amp load, voltage, phase, rpm, service factor and other data. Record and compare with the design requirements written on the equipment sheets. If there are discrepancies in the voltage, phase or rpm they must be reconciled.

2. *Thermal Overloads.* Go to the starter next and check that the thermal overloads are installed and that they are the correct size. In a 3-phase system there must be three overloads, one for each line. If they are not installed, do not test the system until they are!

   The thermal overloads must also be the correct size and not exceed the motor nameplate amps. For example if the maximum nameplate amps are *12.0,* the thermal overload must be rated for a maximum of *12* amps, plus or minus a few tenths. The correct size overload is normally on a chart on the inside of the starter cover. Locate the maximum amps in the column and read the size heater required next to it. Usually the heater number is stamped on the face of the heater itself and is visible when installed.

Inspect Fan Components

1. *Fan Wheel.* Inspect the fan wheel next. Is it the correct type and size? On centrifugals it could be one of four basic types, backward inclined, air foil, forward curve or paddle wheel.

   Is the fan wheel installed correctly? Sometimes the factory installs a fan wheel backwards in a fan, or if the fan is knocked down and assembled on the job site it frequently will be installed backwards.

Is the *gap and center line alignment* between the wheel and the inlet cone on centrifugal fans correct? This can cause internal fan cycling and major havoc on the fan performance, reducing air flow 30, 40 or 50%.

Check to see that the wheel *is securely fastened* to the shaft. Check that the bearings are greased properly if they are not the permanently lubricated type.

2. *Drives.* Inspect the drives. Is the belt *tension* correct? On multi-belted drives is the tension the same on each belt? If not, it could indicate that the belts are of different lengths and are not a matched set.

   Is the *alignment* correct? Cockeyed belts wear out fast and do not efficiently transmit horsepower.

   Make a rough mental calculation of the pulley diameter *ratio* and compare with the motor/fan rpm ratio. Catastrophes have occurred when new or remodeled systems were first turned on. Ducts and plenums have burst apart or collapsed due to incorrect pulley ratios.

$$\frac{\text{Fan Pulley Dia.}}{\text{Motor Pulley Dia.}} = \frac{\text{Design Motor rpm}}{\text{Design Fan rpm}}$$

   Example:

$$\frac{10'' \text{ Dia.}}{5'' \text{ Dia.}} = \frac{1800 \text{ rpm}}{900 \text{ rpm}} = 2$$

   Record pulley diameters, belt sizes, the true center distance from the motor shaft to the fan shaft and available motor movement back and forth.

3. *Bump* the fan to check the rotation of the wheel. Frequently motors are wired in reverse. To reverse the direction of a three-phase motor, switch two leads at the motor or starter. For single-phase starters check the motor wiring diagram. Bumping the fan simply means turn the fan on and off again quickly.

Check System Components

1. Inspect the *filters* to see that they are installed and clean. On new jobs, if they are a temporary construction set, replace with the permanent set. If a permanent set, make sure they are not excessively dirty or clogged.

2. *Cooling and heating coils.* Check the cooling and heating coils. In built up housings; are they properly blanked off all around the tops and bottoms and sides so air does not

bypass the coil? Are there large gaps where the piping connections protrude through the side of the housing? If so, seal properly

Check if the coils are clean. If the system must be balanced with the heating or cooling on, are the coils and control valves in proper operation? If balancing must be done in a cooling mode and the cooling system is not operable for whatever reason, portions of the coil face area can be blocked off with cardboard or polyethylene to assimilate a wet coil pressure drop.

3. Automatic Dampers. The next step in the system component checkout is to check and set the automatic dampers in their balancing positions. There are two approaches in settings of outside air, return air and exhaust control dampers.

The first approach, if there is a *separate* RA fan, is to set the outside air to 100 percent closed, and the exhaust dampers to 100 percent open. After balancing on 100 percent OA set the OA to minimum and spot check outlets and fan discharge flow.

If weather conditions prohibit 100% OA, set the OA to the minimum position and the return air in its maximum and then balance. This puts the maximum load on the supply fan that it will ever have to handle. If it works under this condition it will also work under any lesser load.

If there is no separate return air fan and the supply fan is handling both the supply and return, the maximum load on the fan is achieved when the OA is at minimum and return air is at maximum. Balance in this mode. Spot checking must be done in the maximum CIA and minimum RA positions.

If there are face and bypass *dampers by* a heating coil, the face damper should be 100 percent open and the bypass closed.

If there are automatically controlled *vortex dampers* on the intake of a centrifugal fan, as with medium and high pressure VAV systems, it should closed completely and then upon start-up opened slowly, to prevent possible bursting of ductwork.

4. *Outlet and ductwork dampers.* After the central equipment is set up, go through the spaces served by the system and shine a flashlight through all outlets to make sure that all the grille and ceiling diffuser dampers are 100 percent open before turning the system on.

Check that *splitter dampers* are positioned at roughly a 30 to 45 degree angle and that other *manual volume dampers* and fire *dampers* are 100 percent open.

5. *Thermostat settings.* On low pressure constant volume single zone systems (a) In winter leave the thermostat on its normal setting. (b) In summer if the cooling is in operation,

set stat to maximum cooling, usually 55 degrees, so that the coil is wetted and the system is balanced under its maximum load.

Take Fan Readings

1. *Start-up.* After completing the inspection and set up of the equipment and dampers, *turn on the system* to be balanced, plus all other systems that serve the same area, and take start-up readings.

   Upon start-up listen for bursting or collapsing ducts, a rubbing fan wheel, motor or bearing noises or rumbling or clanging of any type. Observe the operation of the automatic dampers. If something erratic is seen or heard, turn off the fan immediately, check out and rectify the problem before proceeding.

2. *Amp and Volt Reading.* Since a motor can be burned up so quickly, the first thing to do after starting the equipment is to check the amp draw, to make sure it is not exceeding maximum motor amps, and check the voltage to confirm it is in the correct range.

   This is normally done with a volt-ammeter at the starter. The jaws are clamped around each wire, one at a time, and the amps read. Then the probes are used to read voltages across terminals.

   If there is a big difference between the amps on the legs, or if the voltage deviates greatly from design or fluctuates, there may be electrical system problems which have to be resolved before you can proceed in testing the system.

3. *Rpm reading.* Immediately after the amp-volt reading, check the fan rpm to see that it is approximately as per the design. Use a tachometer.

   If the rpm of the fan is grossly higher or lower than design check the following:

   a) Check the motor rpm to see if a wrong speed motor was installed.

   b) Check the pulley diameters to see if you have the correct diameter ratio.

   c) Check if the blueprints, fan drawings or test report sheets are in error, or if there was a change.

   The drive belts may also simply be riding too high or low in variable pitch motor pulley. If the amp draw permits it, change the variable pitch sheave to get the fan at the correct rpm.

4. *Fan suction and discharge pressures.* Read the fan suction and discharge static pressures next and add them together for the total fan static pressure. For example, a typical

suction pressure may be 1 inch and the discharge .5 inches. This would be a total of 1.5 inches.

5. *Pressure drops across suction side components.* The pressure drops across the filters and coils should be taken next for possible flow problems analysis and future reference. They also serve as a check against the design engineer's calculations and equipment manufacturers' catalogue ratings.

   The *pressure drops* across the filters, coils and control dampers can be taken with a magnehelic gauge with a 0- to 1-inch or 0- to 2-inch scale. Drill holes in the component housing on entering and leaving sides of components. Take individual static pressure readings at each point and subtract upstream from downstream readings for arriving at the drops.

6. *Total Air Flow.* Knowing three characteristics of the fan performance out of the five, the amp draw, rpm and fan static pressure, the fourth critical aspect is checked at this point.

   Check the total air flow from the supply fan to see if you have approximately the correct amount to start off with, before balancing the outlets.

   Exceptions to this procedure are where there are no duct sections for accurate total flow readings, and for small systems with few outlets, where it is easier and faster to read all the outlets and total them up than to take a Pitot traverse at the fan.

   The most accurate method of taking a total air flow reading is with a *Pitot tube traverse in* a straight run of ductwork five to ten times the width of the duct. Readings must *not* be taken in or near fittings or after dampers, coils, and so on, because of the potential turbulent flow and unreliability of readings at these points. Enough points in a duct cross section must be read for a valid velocity average.

   If a Pitot tube traverse cannot be taken in the main discharge duct due to fittings, equipment, lack of straight duct, inaccessibility, etc., traverse readings with an anemometer can be taken on the discharge side of a filter or coil. These readings usually are not very accurate, but they will provide a rough idea of the total cfm in order to determine if the fan is running all right and if balancing the outlets is feasible.

7. *Stratification Check.* Air stratification through coils, filters, louvers, dampers, etc. can cause coil freeze up, under or over heating or cooling, and great energy inefficiency. If the arrangement of the outside air flow and return air flow into the mixing plenum gives any indication that there might be poor mixing of the air resulting in temperature or velocity stratification, check out for stratification.

Proportionate Balance Outlets and Duct Runs

After the equipment is found to be correct and the total cfm in the right range, the outlets and branch ducts can be balanced.

First walk through the various areas served by the system to see if there are any problems with temperatures, drafts, air noises, etc. Spot check some end, middle and starting outlets and duct runs to roughly determine the extent of imbalance. Then proceed with the balancing.

The most effective method of balancing is the "proportionate" method. This method results in the least amount of energy usage by the fan. In proportionate balancing, all outlets and branch ducts in the system, starting with those farthest from the fan, are brought to about the same percent of design, give or take 5 percent. The method of doing this will be covered in the following chapter.

Constant volume, single zone, low pressure supply systems and exhaust systems can be proportionately balanced. High pressure systems involve a slightly different procedure for the high pressure side of the system.

Final Settings and Readings at Fan

After the outlets and branch ducts are proportionately balanced (that is, they are all approximately at the same percent of design, whatever it may be, 85, 95, 100 or 115 percent), return to the fan and recheck the total cfm, amps and discharge static pressure.

If the total of the outlet cfm is ten percent or more higher or lower than design, the fan rpm should be increased or decreased to get as close to 100% of design as possible.

If the outlets were correctly proportionately balanced, their flows will all increase or decrease roughly the same percentage as the fan cfm. For example, if the fan is increased 12 percent the outlets will each do likewise. Or if the fan flow is decreased 15 percent the outlets will also decrease.

If the cfm *must be increased,* check the actual amps against the motor full rated amps to make sure they will not exceed the higher fan rpm and air flow. Check to see if there is any room on the variable pitch drive on the motor to alter the rpm.

Using *fan law number 1* calculate the *new rpm* needed to achieve the new cfm. Calculate also what the *new static pressure* and *break horsepower* are. Compare bhp with actual hp of motor. Determine if you need new belts or sheaves. Check *motor movement* forward and backward in regard to whether belts can be reused or not. If a new motor is indicated consider if you live with less cfm to retain existing motor.

After changing the cfm at the fan, spot check key outlets in each branch to verify they have increased or decreased proportionately.

Unsealed low pressure ductwork might leak from 5 to 15 percent of the air flow. The average is about 8 percent. The fan cfm, under these conditions of leakage will not match the total flow at the outlets and will generally run about 8 percent less.

Medium and high pressure ductwork which has been properly sealed and leak tested should not leak more than 1 percent. The fan and total outlet cfm should be relatively the same in these situations.

EQUIPMENT TEST REPORTS

Detailed test reports for each major piece of equipment in the building are very important in balancing.

These reports record the key performance figures such as:

— gpm
— cfm
— Amp draws
— Pressures
— Temperatures, etc.

They also record important names such as:

— Detailed description of the equipment
— Manufacturers—Model numbers
— Maximum amp draws on motors
— Service factors
— Settings of dampers and valves
— Pressure drops

Test reports are required on major equipment such as fans, pumps, chillers, condensers, boilers, etc.

PROCEDURE DIAGRAM FOR TESTING AND BALANCING AN HVAC SYSTEM

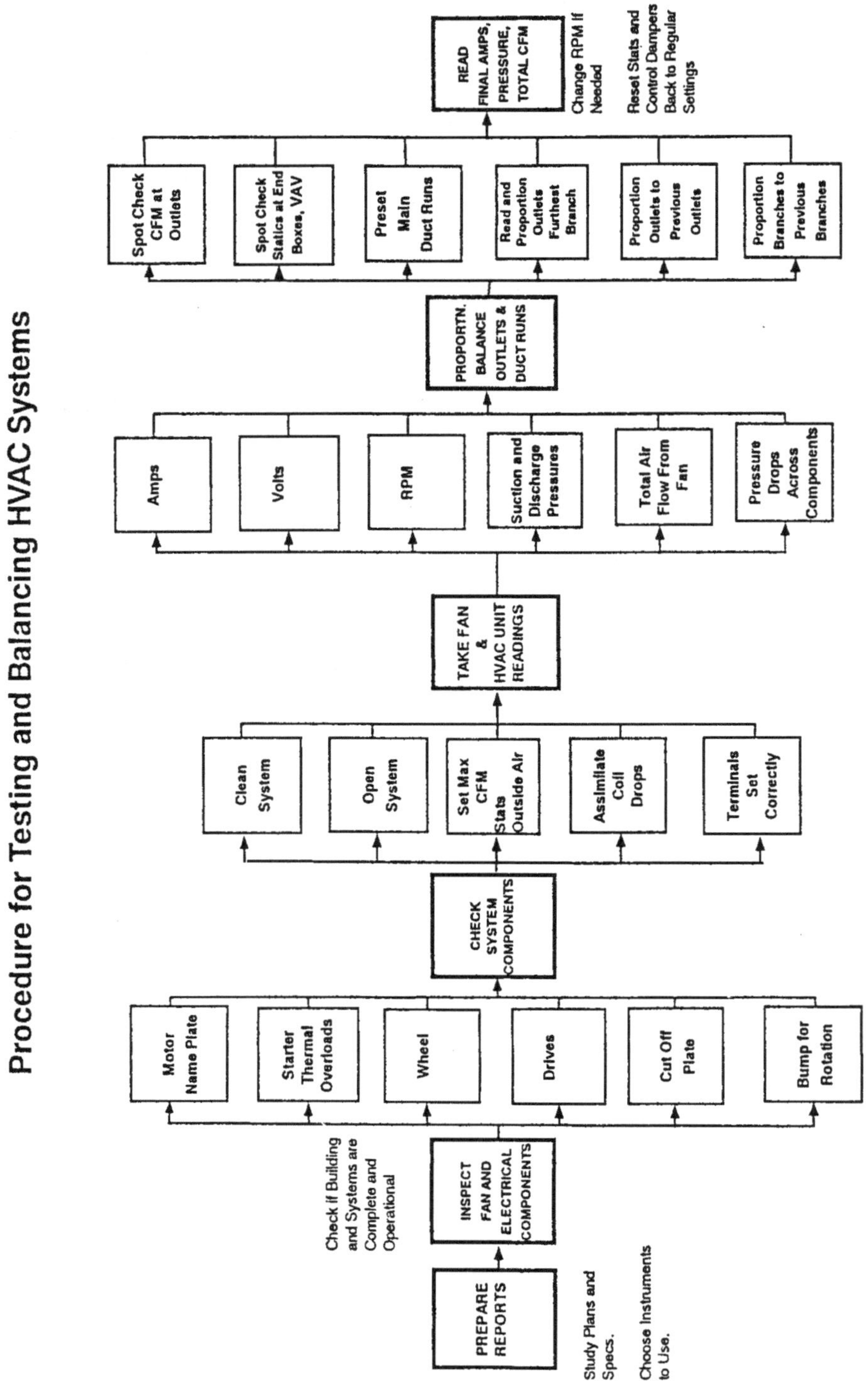

Figure 1-1. Procedure for Testing and Balancing an HVAC System.

TESTING AND BALANCING REPORT

Date ____________________

Job ____________________

Location ____________________

Architect ____________________ Phone ____________________

Engineer ____________________ Phone ____________________

Testing and Balancing Contractor ____________________

Address ____________________

City ____________________ State ________ Zip ____________

Phone ____________________ Fax Number ____________________

Figure 1-2. Sample Filled Out Testing and Balancing Audit Report Cover Sheet

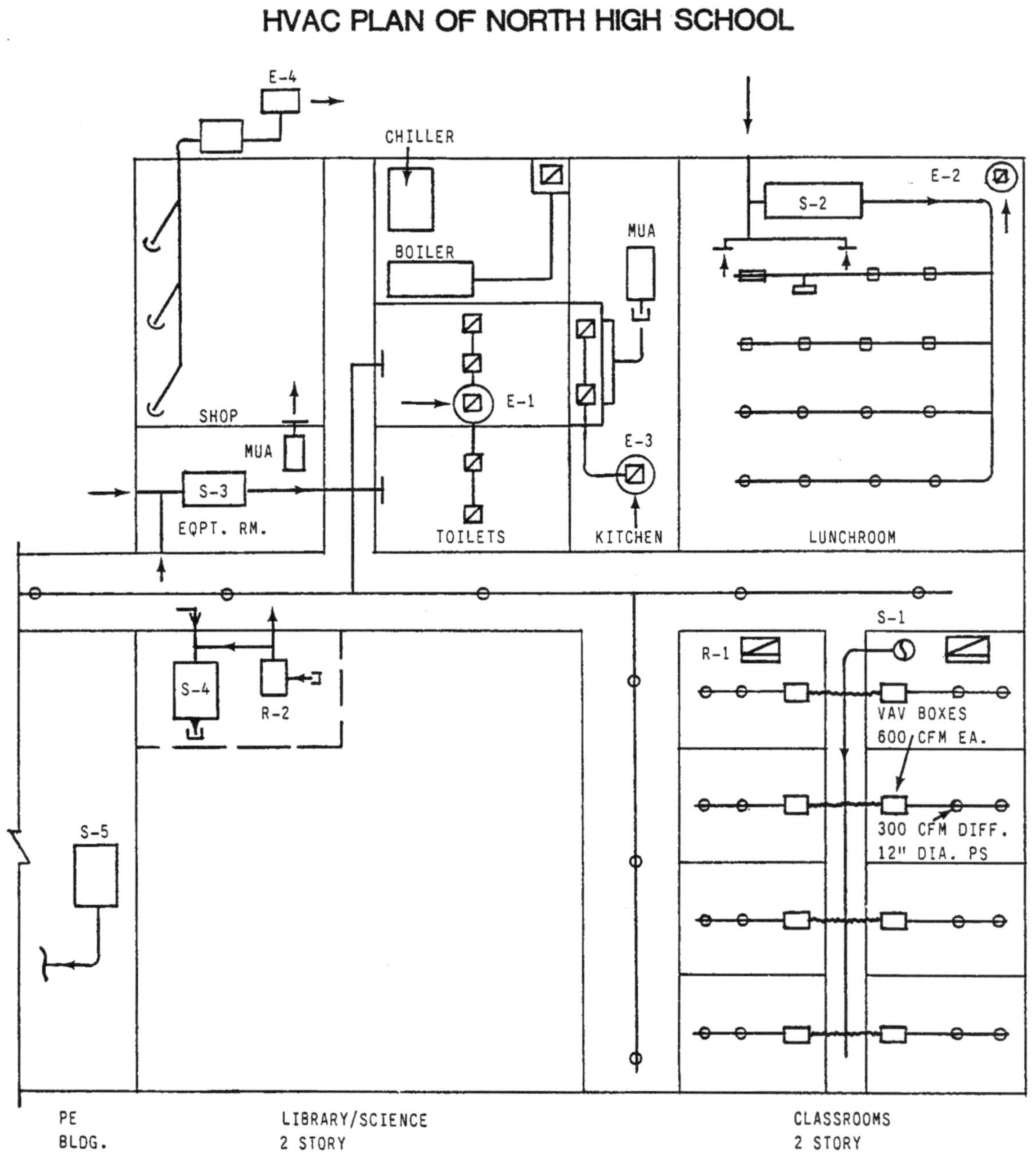

Figure 1-3. Diagrammatic plan of North High School Used for Sample Forms which Follow.

GENERAL INFORMATION

Job North High School Job No. C-150 Date Aug. 15, 1981

Location

TYPE SYSTEMS

Low pressure single zone — Chilled water
Medium pressure VAV — Hot water
Air handling units
Roof top units
Make up air units
Toilet exhaust
Kitchen exhaust
Dust collector system

DIFFUSERS AND REGISTERS

Manufacturer Tuttle and Bailey

Types Round diffusers fixed, PS — Linears 4000
Round diffusers adjustable, PA
Registers, 1100

TERMINAL UNITS

Manufacturer Tuttle and Bailey

Types VAV boxes, pressure independent, pneumatic, normally closed

INSTRUMENTS USED (Indicate Models)

- ☑ Velometer
- ☑ Anemometer
- ☑ Manometers
- ☑ Volt-Ammeter
- ☑ Flow Hood
- ☐ Thermal Anemometer
- ☑ Magnehelics
- ☑ Tachometer

BUILDING AND SYSTEM COMPLETION CHECK OFF LIST

Architectural: ☑ Walls ☑ Roof ☑ Floors ☑ Windows ☑ Doors ☑ Ceilings

Electrical: ☑ Starters ☑ Overload ☐ Transformers Tested ☑ Wiring

Controls: ☑ Control Motors ☑ Linkages ☐ Compressors ☑ Stats ☑ Tubing ☑ Wiring

Piping: ☑ Coils ☑ Valves ☑ Piping ☑ Pumps ☑ Wiring

Sheet Metal: ☑ Grilles ☑ Fins ☑ Drives ☑ Air Handling Units ☑ Filters ☑ Wiring

Remarks

Figure 1-4. Sample Filled Out General Information for Testing and Balancing Form

FAN TEST REPORTS

Fan test reports make up one half of the critical work done in balancing. The fans are the heart of the system and must be running properly before duct runs and outlets are balanced.

1. Fill out data on fan, motor and starters from drawings.

2. Check *nameplate* information and make physical inspection for further data and enter.

3. Check starters and overload sizes.

4. Bump fan for correct rotation.

5. Finally, take electrical, rpm, pressure and flow *readings* and enter.

Fan cfm and the total cfm at the outlets might not match if there is ductwork leakage. Actual amps at motor must not exceed rated amps. Compare fan readings such as rpm, cfm and static pressure-to-fan *charts*.

FAN TEST REPORT

Job North High School Job No C-150 Date Aug. 1, 1981

Location ____ System S-2

Equipment Location Mezzanine Serves Lunchroom Tested By: HW

☑ Air Handling Unit ☐ Roof Top Unit ☐ Furnace ☐ Supply Fan ☐ Exhaust Fan ☐ Pkg Unit

☑ LP ☐ MP ☐ HP ☑ Constant Volume ☐ VAV

FAN DATA	
Manufacturer	Barry
Model Size	AF 7245 DWDI
Type Fan	☑ Centrigal ☐ Roof Exhaust ☐ Inline ☐ Vane Axial ☐ Prop.
Type Wheel	☐ Backward Incline ☑ Air Foil ☐ Forward Curve ☐ Paddle Wheel
Wheel:	☑ Alignment OK ☑ Gap ☑ Fastened ☑ Clean
Belts (2) B131	C to C Distance 52"
Pulleys: Fan Dia. 10"	Mot. Dia. 5"
Motor Movement 2" ±	Belts
Bearings ✓	☑ Cut Off Plate OK

MOTOR		
Manufacturer GE	Serial No.	
Frame No. 184T	Type Frame ☐ T ☑ U	
Service Factor: 1.15	Rated	Actual
HP, Nameplate	5	5
BHP $[HPnp \times \frac{Aa}{Ar} \times \frac{Va}{Vr}]$	3.53	3.84
Amps, L_1 L_2 L_3	15.2	12.2
Voltage, L_1 L_2 L_3	230V	220V
RPM	1750	1750
Phase	3	3

STARTER	
Manufacturer GE	Model 141R
Starter Size 0	Class 1
Overload: Required Size	CR 15.4
Actual	CR 15.4

FAN PERFORMANCE	Design	Actual
Fan CFM	9,800	10,160/104%
Outlet CFM Total	9,800	9,756/99%
Fan RPM	985	992
Fan S.P.	1½"	1.6"

STATIC PRESSURE DROPS	Upstream	Downstream	Total Drop
Filter	.2"	.4"	.2" W.G.
Heat. Coil	.4	.6	.2
Cool. Coil	.6	1.0	.4
Fan Inlet			1.0
Fan Discharge			.6
Total Fan S.P.			1.6"

CONDITIONS	
Vortex Damper Position	—
Outside Air Damper Setting	4000 CFM
Return Air Damper Setting	6160 CFM
Filter Conditions	Clean
Coil Conditions	Clean
Temperatures	
OA: @ 40%	40 F DB WB RH
RA:	70 F DB WB RH
Mixed Air:	DB WB RH
Discharge	DB WB RH
Space:	DB WB RH
Duct Temp. Drop	DB

Remarks ____

PROBLEMS:

☐ Too much air ☐ SP Low ☐ Too Hot

☐ Too little air ☐ SP High ☐ Too Cold

☐ Air Noises ☐ Fan Noises

☐ Oversized equipment ☐ Undersized Equipment

☐ Other ____

Wendes Engineering and Contracting Services 7/80 Form TAB 201

Figure 1-5. Sample Filled Out Fan Test Report Form

OUTLET AIR BALANCE REPORT

Air distribution systems must be planned, checked out and balanced in an organized manner. The outlet balance reports initiate and control this process.

List the area being served, the *number* of the outlet, *model, size* and *cfm* required. The Ak (effective area of flow through the outlet) and required velocity are not needed if a flow *hood* (which reads cfm directly) is being used.

If air velocities are being read with an Alnor Velometer, Ak areas have to be looked up in manufacturer's factor manual or determined by some other means and entered. Also, the required velocity and required cfm must be calculated from the Ak.

Notes on the right-hand side of the outlet air balance report refer to the branch ducts in the air distribution system, B1, B2, etc., and to the percentage of design of each branch after each pass in the proportionate balancing process.

OUTLET AIR BALANCE REPORT

Project North High School Job No. C-150 Date Aug. 1, 1981

Location Lunchroom System S-2

Instruments Used Velometer Tested by: HW

ROOM AREA SERVED	OPENING No.	OPENING Model	OPENING Size	OPENING A_k	REQUIRED Vel	REQUIRED CFM	PRELIMINARY Percent of Design			FINAL Vel	FINAL CFM	
Lunchroom Student Area	1	PS	12" ϕ	.66	909	600	70%	80/85	90/100		600	B1 90% 95% 100% Avg
	2		12" ϕ	.66	909	600	90%	80/85	92/102		612	
	3		12" ϕ	.66	909	600	95%	85	87/97		582	
	4		12" ϕ	.66	909	600	105%	90	100		600	
	5	PA	14" ϕ	.78	1025	800	80%	85/90	102		816	B2 100% 95% 100% Avg
	6		14" ϕ	.78	1025	800	90%	85/90	98		784	
	7		14" ϕ	.78	1025	800	105%	90	99		792	
	8		14" ϕ	.78	1025	800	130%	101			808	
	9	DF	2415	.65	1076	700	95%	105	106/96		672	B3 110% 100% Avg
	10		2415	.65	1076	700	95%	105	110/100		700	
	11		2415	.65	1076	700	125%	105	113/103		721	
	12		2415	.65	1076	700	125%	111	101		707	
Teacher Area	13	4000 Linear	Size 3 6 Ft	.11×6 = .65	756	500	80%	85/90	98		490	B4 97%
	14	Dual Troffer	4 Ft	.15	666	100	90%	85/90	96		96	
0° Deflect.	15	T57 Reg	12×12	.71	563	400	100%	90	97		388	
40° Deflect.	16	T57 Reg	12×12	.57	701	400	130%	97			388	
			Total			9,800	CFM				9,756	Avg. 99%
Lunchrm RA	1	T70D	48×24	6.8	441	3000					3100	
	2	T70D	48×24	6.8	441	3000					3060	
			Total			6000	CFM				6160	

Remarks

Mendes Engineering and Contracting Services Form No. TAB202

Figure 1-6. Sample Filled Out Outlet Air Balance Report Form

AIR SYSTEMS RECAP

The air system recap keeps critical fan and motor readings in a overall list for easy checks of multi-system projects and for keeping track of the status of work.

The critical fan and motor readings, cfm, rpm, S.P. and AMPS, plus the status of balancing, are recorded on this form.

AIR SYSTEMS RECAP

Job North High School Job No. C-150 Date Aug. 15, 1981

Location ______ Balancer HW

SYSTEM	LOCATION OF EQUIPMENT	STATUS 1	STATUS 2	STATUS 3	CFM Design	CFM Actual	CFM Percent of Design	RPM Design	RPM Act.	S.P. Inches Des.	S.P. Inches Act.	AMPS Des.	AMPS Act.	REMARKS
S-1	Classrm. Penth.	✓	✓	✓	10,400	10,600	102%	1400	1410	4	3.8	15.2	13.5	AHU
S-2	Lunchroom	✓	✓	✓	9,800	9,756	99%	985	992	1½	1.6	15.2	12.2	AHU
S-3	Eqpt. Rm. 110	✓	✓		17,000	13,600	80%	900	880	2	1.8	28.0	21.7	AHU
S-4	Library	✓			17,000			900		2		28.0		AHU
S-5	Gym	✓			22,000			895		2½		42.0		AHU
R-1	Classrm. Penth	✓			8,000			897		1		9.6		Centrifugal
R-2	Library				13,000			780		1½		15.2		Centrifugal
E-1	Toilets				3,200			740		½		2.0		Roof Exh Fan
E-2	Lunchroom	✓	✓		2,500	2,450	98%	751	750	½	½	2.0		Roof Exh Fan
E-3	Kitchen				5,400			789		1		6.8		Roof Exh Fan
E-4	Shop				10,000			1252		4½		42.0		Industrial Exh Fan

1. System ready for balancing 2. Equipment checked out 3. System balanced

kendes Engineering and Contracting Services 7/80 Form TAB 206

Figure 1-7. Sample Filled Out Air Systems Recap Form

PITOT TUBE TRAVERSE RECTANGULAR DUCT

To determine the volume of air flow in a duct in terms of cfm, a multi-point Pitot tube traverse is taken inside the duct according to a certain procedure and the readings averaged for the overall cross sectional area of air flow.

1. Write in the width and depth of the duct being traversed.

2. The cross sectional area of duct is divided into equal areas and the center point of each is read and averaged by number of points.

3. Follow instructions on form to determine points of reading. Divide recommended spacing (from bottom chart) into duct width and split remainder in half for distance from sides of duct.

4. Do same for duct height.

5. Read velocities with Pitot tube in either inches of water gauge or directly in feet per minute depending on instrument being used.

6. Add up fpm readings and divide by number of points read for the average velocity in the duct.

7. Multiply times the cross sectional area in sq. ft. to determine cfm of flow.

PITOT TUBE TRAVERSE, RECTANGULAR DUCT

Job North High School Job No. C-150 Date Aug. 1, 1981

Location Lunchroom System S-2

Location of Duct Mezzanine Area Served Lunchrm Duct Temperature 70 F & SP .6"

Duct Size 48 x 24	Required CFM 9,800	Required FPM 1225
Duct Area Sq Ft	Actual CFM 10,160	Actual FPM 1270

☐ Initial ☒ Final

PERCENT OF DESIGN: (Divide actual CFM by required CFM) = (10,160) / (9,800) = 104 percent

	1 VP / FPM	2 VP / FPM	3 VP / FPM	4 VP / FPM	5 VP / FPM	6 VP / FPM	7 VP / FPM	8 VP / FPM	Spacing
1	.06 / 981	.08 / 1133	.10 / 1266	.11 / 1328	.12 / 1387	.13 / 1444			3" (A/2)
2	.07 / 1060	.07 / 1060	.09 / 1201	.11 / 1328	.13 / 1444	.14 / 1498			6" (A)
3	.07 / 1060	.08 / 1133	.10 / 1266	.12 / 1387	.14 / 1498	.12 / 1387			6"
4	.06 / 981	.07 / 1060	.10 / 1266	.13 / 1444	.12 / 1387	.14 / 1498			6"
5									3"
6									A/2
Column Totals	4082	4386	4999	5487	5716	5827			= Total of Columns
Spacing	4" (A/2)	8" (A)	8"	8"	8"	8" 4"		A/2	

A. Divide recommended spacing into duct width for equal spacing between points of readings and write in each space.

B. Split remainder in half for spacing at sides of duct and write in.

C. To determine the actual average FPM, total the FPM in each column, add the columns together and divide by total number of points in traverse.

Total FPM / Total No. of points = (30,497) / (24) = 1270 Average FPM

RECOMMENDED EQUAL SPACING BETWEEN POINTS OF READING	
Width Range of Duct Inches	Spacing in inches
8-14	5
15-24	6
25-36	7
37-48	8
49-60	9
61-72	10
73-84	11
85-96	12

CONVERTING VP INTO FPM

VP	FPM	VP	FPM	VP	FPM	VP	FPM	VP	FPM
.01	400	.31	2230	.61	3127	.91	3821	1.21	4405
.02	566	.32	2260	.62	3153	.92	3842	1.22	4423
.03	693	.33	2301	.63	3179	.93	3863	1.23	4442
.04	801	.34	2335	.64	3204	.94	3880	1.24	4460
.05	895	.35	2369	.65	3229	.95	3904	1.25	4478
.06	981	.36	2403	.66	3254	.96	3924	1.26	4495
.07	1060	.37	2436	.67	3279	.97	3945	1.27	4513
.08	1133	.38	2469	.68	3303	.98	3965	1.28	4531
.09	1201	.39	2501	.69	3327	.99	3985	1.29	4549
.10	1266	.40	2533	.70	3351	1.00	4005	1.30	4566
.11	1328	.41	2563	.71	3375	1.01	4025	1.31	4583
.12	1387	.42	2595	.72	3398	1.02	4045	1.32	4601
.13	1444	.43	2626	.73	3422	1.03	4064	1.33	4619
.14	1498	.44	2656	.74	3445	1.04	4084	1.34	4636
.15	1551	.45	2687	.75	3468	1.05	4103	1.35	4653
.16	1602	.46	2716	.76	3491	1.06	4123	1.36	4671
.17	1651	.47	2746	.77	3514	1.07	4142	1.37	4688
.18	1699	.48	2775	.78	3537	1.08	4162	1.38	4705
.19	1746	.49	2804	.79	3560	1.09	4181	1.39	4722
.20	1791	.50	2832	.80	3582	1.10	4200	1.40	4739
.21	1835	.51	2860	.81	3604	1.11	4219	1.41	4756
.22	1879	.52	2888	.82	3625	1.12	4238	1.42	4773
.23	1921	.53	2915	.83	3657	1.13	4257	1.43	4790
.24	1962	.54	2943	.84	3669	1.14	4276	1.44	4806
.25	2003	.55	2970	.85	3690	1.15	4295	1.45	4823
.26	2042	.56	2997	.86	3709	1.16	4314	1.46	4840
.27	2081	.57	3024	.87	3729	1.17	4332	1.47	4856
.28	2119	.58	3050	.88	3758	1.18	4350	1.48	4871
.29	2157	.59	3076	.89	3779	1.19	4368	1.49	4889
.30	2193	.60	3102	.90	3800	1.20	4386	1.50	4905

Wendes Engineering and Contracting Services 7/80 Form TAB 203

Figure 1-8. Sample Filled Out Pitot Tube Traverse, Rectangular Duct Form

PITOT TUBE TRAVERSE SMALL ROUND DUCTS

Either six or ten horizontal and vertical points of reading are needed across a round duct depending on whether the duct size range is 3" to 10" or over 10."

The points are in the center of equal area concentric circles, and the locations from the side of the duct are predetermined and listed in the chart.

Read velocities in either inches of water gauge or directly in feet per minute depending on instrument used.

Add up fpm readings and divide by the number of readings, 6 or 10, for the average velocity in the duct and multiply times the cross sectional area in sq. ft. to determine cfm of flow.

PITOT TUBE TRAVERSE, SMALL ROUND DUCTS

Job SEARS Job No. C-120 Date 5-17-88

Location CHICAGO System S-2

Location of Duct ______ Area Served 1st Duct Temperature 55° ℄ SP .09"

Duct Diameter 8"	Required CFM 470	
Duct Area Sq Ft	Actual CFM 450	Actual FPM

Percent of Design: = Actual CFM / Required CFM = (457) / (470) = 97 percent

TRAVERSE POINT LAYOUT

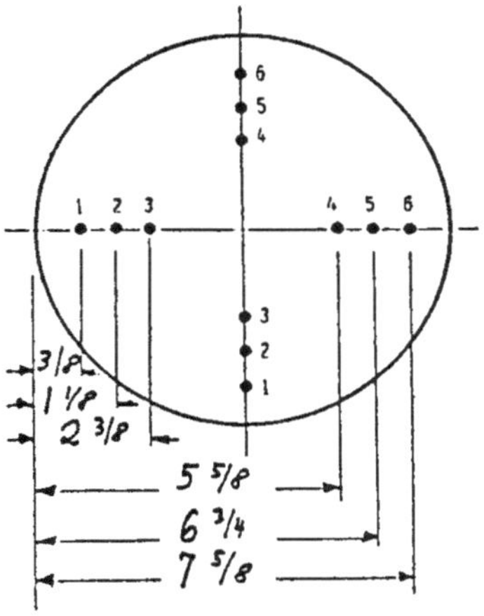

Distance from side of duct to point of reading to nearest 1/8th inch for 6 point traverse.

TRAVERSE POINTS	READINGS At Startup V.P.	FPM	Final V.P.	FPM
1			.11	1328
2			.13	1444
3			.14	1498
4			.09	1201
5			.07	1060
6			.11	1328
1				
2				
3				
4				
5				
6				
Total FPM				7859
Divided by 6 =				1310
Times Duct Area	.348			
= Total CFM				457

Duct Dia.	POINTS 1	2	3	4	5	6
3	1/8	1/2	7/8	2 1/8	2 1/2	2 7/8
4	1/8	5/8	1 1/8	2 7/8	3 3/8	3 7/8
5	1/4	3/4	1 1/2	3 1/2	4 1/4	4 3/4
6	1/4	7/8	1 3/4	4 1/4	5 1/8	5 3/4
7	1/4	1"	2"	5"	6"	6 5/8
8	3/8	1 1/8	2 3/8	5 5/8	6 3/4	7 5/8
9	3/8	1 3/8	2 5/8	6 3/8	7 5/8	8 5/8
10	3/8	1 1/2	3"	7"	8 1/2	9 1/2

CONVERTING VP INTO FPM

VP	FPM	VP	FPM	VP	FPM	VP	FPM	VP	FPM
.01	400	.31	2230	.61	3127	.91	3821	1.21	4405
.02	566	.32	[illegible]	.62	[illegible]	.92	3842	1.22	4423
.03	693	.33	2301	.63	[illegible]	.93	3863	1.23	4442
.04	801	.34	2335	.64	3203	.94	3883	1.24	4460
.05	895	.35	[illegible]	.65	[illegible]	.95	[illegible]	1.25	4478
.06	[illegible]	.36	2403	.66	3254	.96	3924	1.26	4495
.07	[illegible]	.37	[illegible]	.67	3279	.97	3945	1.27	4513
.08	1133	.38	[illegible]	.68	3301	.98	3965	1.28	4531
.09	1201	.39	[illegible]	.69	3327	.99	3985	1.29	4549
.10	[illegible]	.40	[illegible]	.70	3351	1.00	4005	1.30	4566
.11	1328	.41	2563	.71	3375	1.01	4025	1.31	4583
.12	1387	.42	[illegible]	.72	3398	1.02	4045	1.32	4601
.13	1444	.43	2626	.73	[illegible]	1.03	4065	1.33	4619
.14	1498	.44	2656	.74	3445	1.04	4084	1.34	4636
.15	1551	.45	2687	.75	3468	1.05	4103	1.35	4653
.16	1602	.46	2716	.76	3491	1.06	4123	1.36	4671
.17	1651	.47	2746	.77	3514	1.07	4142	1.37	4688
.18	1699	.48	2775	.78	3537	1.08	4162	1.38	4705
.19	1746	.49	2804	.79	3560	1.09	4181	1.39	4722
.20	1791	.50	2832	.80	3582	1.10	4200	1.40	4739
.21	1835	.51	2860	.81	3604	1.11	4219	1.41	4756
.22	1879	.52	2888	.82	3625	1.12	4238	1.42	4773
.23	1921	.53	2916	.83	3657	1.13	4257	1.43	4790
.24	1962	.54	2941	.84	3669	1.14	4276	1.44	4806
.25	2003	.55	2970	.85	3690	1.15	4295	1.45	4823
.26	2042	.56	[illegible]	.86	3709	1.16	[illegible]	1.46	4839
.27	2081	.57	3024	.87	3729	1.17	[illegible]	1.47	4855
.28	2119	.58	3050	.88	3758	1.18	[illegible]	1.48	4871
.29	2157	.59	3076	.89	3779	1.19	[illegible]	1.49	[illegible]
.30	2193	.60	3102	.90	3800	1.20	4386	1.50	4905

Mendes Engineering and Contracting Services 7/80 Form TAB 204a

Figure 1-9. Sample Filled Out Pitot Tube Traverse, Small Round Ducts Form

PITOT TUBE TRAVERSE LARGE ROUND DUCTS

It is very critical that Pitot tube traverse points for round ducts be accurately determined.

Traverses of round ducts follow the same principle as those of rectangular ducts in that readings are taken in the center of equal areas.

The difference with round ducts is that you are working with concentric rings of equal area and the rings are narrower on the periphery of the duct and wider towards the center. They are not as easily visualized or calculated as with rectangular ducts, and a chart is needed to establish the distance the Pitot tube is inserted into the duct for each reading.

Both horizontal and vertical traverses are required.

As in the example of the 28-inch-diameter duct, the first reading is 3/4," the second

2 1/4" and the last point is 27 1/4" in which is 3/4" from the opposite side of the duct.

PITOT TUBE TRAVERSE, LARGE ROUND DUCTS

Job North High School Job No. C-150 Date Aug. 1, 1981

Location VAV, MP System S-1

Location of Duct Penthouse Area Served ______ Duct Temperature ______ ℄ SP ______

Duct Diameter 28"	Required CFM 10,400	Required FPM 2,432
Duct Area 4.267 Sq Ft	Actual CFM 10,195	Actual FPM

Percent of Design: $= \frac{\text{Actual CFM}}{\text{Required CFM}} = \frac{(10,195)}{(10,400)} =$ 98 %

TRAVERSE POINT LAYOUT

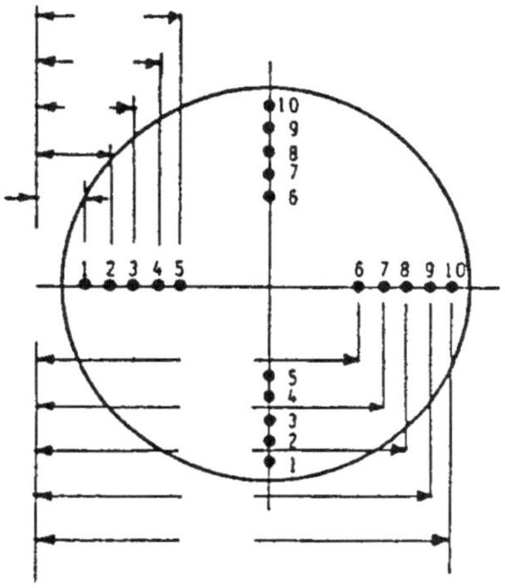

Distance from side of duct to point of reading to nearest 1/8th inch for 10 point traverse.

DUCT DIA.	1	2	3	4	5	6	7	8	9	10
12	3/8	1	1-3/4	2-3/4	4-1/8	7-7/8	9-1/4	10-1/4	11	11-5/8
13	3/8	1	1-7/8	2-7/8	4-1/2	8-1/2	10-1/8	11-1/8	12	12-5/8
14	3/8	1-1/8	2	3-1/8	4-3/4	9-1/4	10-7/8	12	12-7/8	13-5/8
15	3/8	1-1/4	2-1/4	3-3/8	5-1/8	9-7/8	11-5/8	12-3/4	13-3/4	14-5/8
16	3/8	1-1/4	2-3/8	3-5/8	5-1/2	10-1/2	12-3/8	13-5/8	14-3/4	15-5/8
17	1/2	1-3/8	2-1/2	3-7/8	5-3/4	11-1/4	13-1/8	14-1/2	15-5/8	16-1/2
18	1/2	1-1/2	2-5/8	4-1/8	6-1/8	11-7/8	13-7/8	15-3/8	16-1/2	17-1/2
19	1/2	1-1/2	2-3/4	4-1/4	6-1/2	12-1/2	14-3/4	16-1/4	17-1/2	18-1/2
20	1/2	1-5/8	2-7/8	4-1/2	6-7/8	13-1/8	15-1/2	17-1/8	18-3/8	19-1/2
22	5/8	1-3/4	3-1/4	5	7-1/2	14-1/2	17	18-3/4	20-1/4	21-3/8
24	5/8	2	3-1/2	5-1/2	8-1/4	15-3/4	18-1/2	20-1/2	22	23-3/8
26	5/8	2-1/8	3-3/4	5-7/8	8-7/8	17-1/8	20-1/8	22-1/4	23-7/8	25-3/8
28	3/4	2-1/4	4-1/8	6-3/8	9-5/8	18-3/8	21-5/8	23-7/8	25-3/4	27-1/4
30	3/4	2-1/2	4-3/8	6-3/4	10-1/4	19-3/4	23-1/4	25-5/8	27-1/2	29-1/4
32	7/8	2-5/8	4-5/8	7-1/4	11	21	24-3/4	27-3/8	29-3/8	31-1/8
34	7/8	2-3/4	5	7-3/4	11-5/8	22-3/8	26-1/4	29	31-1/4	33-1/8
36	1	3	5-1/4	8-1/8	12-3/8	23-5/8	27-7/8	30-3/4	33	35
38	1	3-1/8	5-1/2	8-5/8	13	25	29-3/8	32-1/2	34-7/8	37
40	1	3-1/4	5-7/8	9	13-5/8	26-3/8	31	34-1/8	36-3/4	39
42	1-1/8	3-3/8	6-1/8	9-1/2	14-3/8	27-5/8	32-1/2	35-7/8	38-5/8	40-7/8
44	1-1/8	3-5/8	6-3/8	10	15	29	34	37-5/8	40-3/8	42-7/8
46	1-1/4	3-3/4	6-3/4	10-3/8	15-3/4	30-1/4	35-5/8	39-1/4	42-1/4	44-3/4
48	1-1/4	4	7	10-7/8	16-3/8	31-5/8	37-1/8	41	44	46-3/4

TRAVERSE POINTS	READINGS At Startup V.P.	At Startup FPM	Final V.P.	Final FPM
1	.22	1879		
2	.30	2193		
3	.40	2523		
4	.55	2970		
5	.41	2563		
6	.36	2403		
7	.36	2403		
8	.37	2436		
9	.36	2403		
10	.28	2119		
1	.28	2119		
2	.22	1879		
3	.36	2403		
4	.30	2193		
5	.37	2436		
6	.40	2523		
7	.55	2970		
8	.41	2563		
9	.36	2403		
10	.36	2403		
Total FPM		47,784		
Divided by 20 =		2389		
Times Duct Area		4267		
= Total CFM		10,195		

Wendes Engineering and Contracting Services

Form TAB 204b

Figure 1-10. Sample Filled Out Pitot Tube Traverse, Large Round Ducts Form

HYDRONIC BALANCE TEST REPORTS

The necessity of working with test report forms in hydronic balancing is just as great as it is in air balancing. Good organization, an outline of information needed, a guide to the sequence of work, readability, and good records are all equally required.

Also flow diagrams are generally required because of the frequent difficulty of following piping systems on blueprints when balancing.

PROCEDURE FOR PREPARING REPORTS

1. Gather required information:
   — Blueprints
   — Shop Drawings
   — Specifications
   — Submittals on:
     • Pumps
     • Chillers
     • HVAC Units
     • Coils
     • Valves
     • Cooling Towers

2. *Study plans and specifications* to become familiar with the types of systems and equipment. Study the routing of the piping and note valve and coil locations. Determine what specifications call for in the way of balancing.

3. Plan the *method* of balancing and select the *instruments* to be used.

4. Prepare flow *diagram* if there is none available. Note the central equipment, terminals, piping, diameters, valves, etc.

5. Fill out *pump test report.* Fill in standard information in heading.

   Enter pump data, manufacturer, model, size, type and impeller size.

   Enter design performance figures, gpm, rpm and full and no flow heads. The suction and discharge pressure readings at full and no flow will be filled in at the job site.

   Fill in the motor data, manufacturer, serial number, mounting frame number, type of internal winding frame, service factor and rated hp, bhp, voltage, rpm and phase. The rated amps will be gotten from the nameplate at the job site as well as from the actual readings.

Enter starter data, manufacturer, model, size and class. The required overload size will be gotten from the inside of the cover on the starter and from the actual overload installed, by inspection.

6. Fill out the flow or *pressure drop water balance report.* List primary and secondary circuits in groups and in sequence. List valves and terminals in the sequence they occur starting at the pump, along with identification and location if needed. Indicate the size. Enter the required cfm and the differential pressure reading required for the particular flow measuring device being used or from the manufacturer's published pressure drop across the item at design flow. List bypasses.

   When proportionately balancing terminals, list percents of design in preliminary readings and gpm only for the final.

7. Fill out *temperature water balance report* for systems without flow measuring stations at terminals, which are generally reheat coils, induction units and baseboard radiation units. List the coil identification and room number in sequence, from the pump out; list primary circuits and secondary circuits as separate groups. Indicate sizes, design entering and leaving water temperature, and entering and leaving air temperatures.

8. If there are chillers fill out a *chiller test report.* Start with standard heating, then the basic compressor data, manufacturer, model, size, type, capacity, refrigerant, pounds and serial number.

   The Freon pressures and temperatures on the compressor, evaporator and condenser are normally checked out by the manufacturer in the start-up and are not generally part of the water balance.

   Fill in design water pressures, temperatures and flows on the *evaporator* and *condenser.*

   Fill in design electrical data, M kW, voltage, phase, etc., on the compressor and starter.

   The conditions at test time will be filled in at the job site—refrigerant and oil levels, water control settings, temperature and pressure cutouts and purge operation.

9. If there are *air cooled condensers and compressors* fill out the appropriate report. Fill in head data, then compressor data, manufacturer, model, size, type, capacity, refrigerant, pounds and serial number.

Fill in design pressures and temperatures on the suction and discharge side of compressor. Enter compressor motor and starter electrical data.

Fill in condenser design temperatures and pressures for liquid line and air. Indicate condenser fan hp, amps and volts.

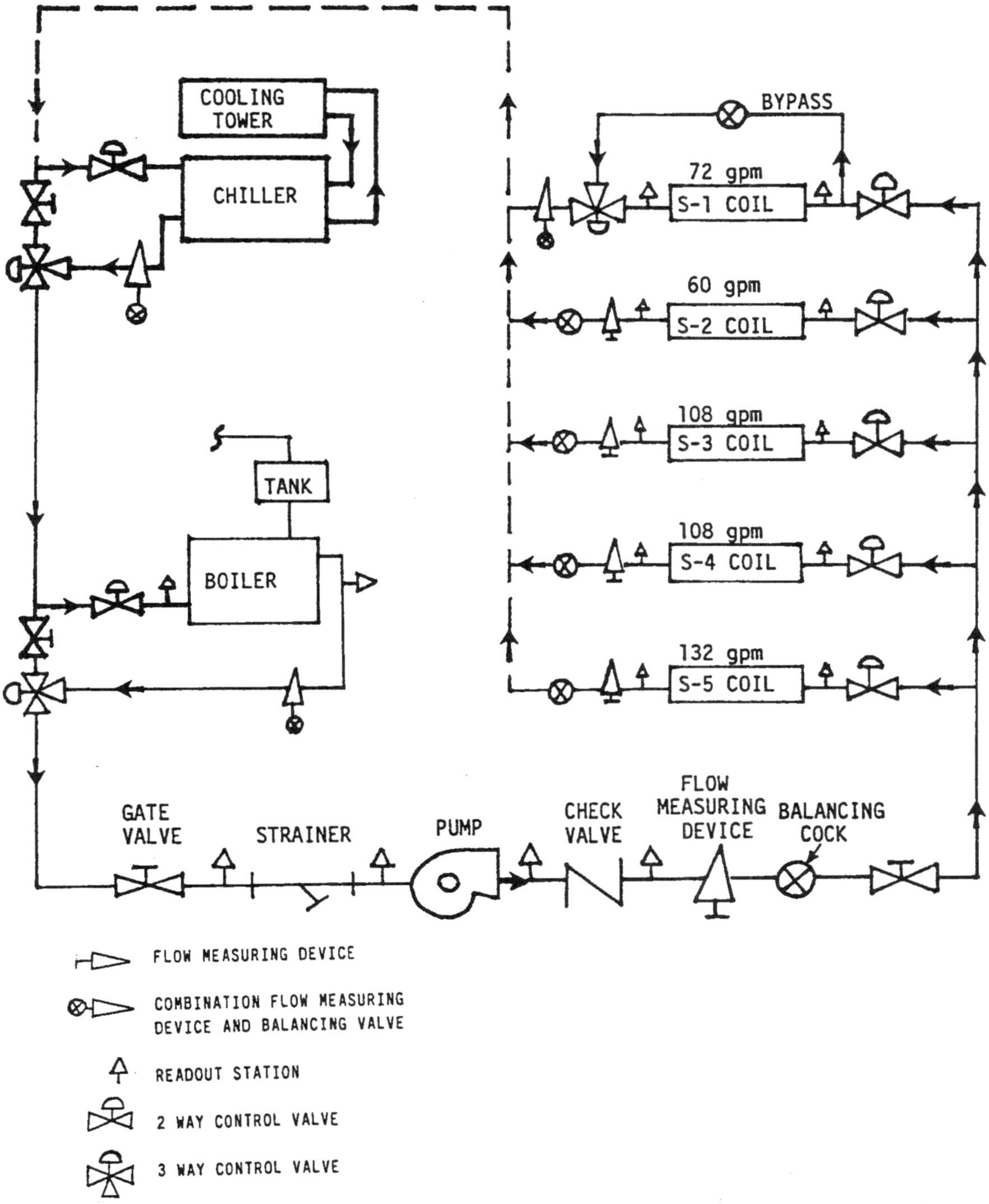

Figure 1-11. Typical two pipe heating and cooling system used for sample test reports.

PUMP TEST REPORT

1. Fill out the heading on the form.

2. Fill out data on pump, motor and starter from drawings.

3. Enter design performance figures, gpm, rpm and full and no flow heads. The suction and discharge pressure readings at full and no flow will be filled in at the job site.

4. Check motor and pump *nameplates* and make physical inspection for further data and enter.

5. Check starter and overload sizes.

6. *Bump* pump for correct rotation.

7. Take electrical readings.

8. Finally, take discharge and static head pressure readings with pump off, then running with no flow, to establish impeller diameter, and then at full flow. Suction pressure is subtracted from discharge for total pressure.

PUMP TEST REPORT

Job North High School Job No. C-150 Date Aug. 15, 1983

Location ____________ System ____________

Equipment Location Penthouse Serves Bldgs A, B, C Tested by: HW

PUMP DATA		
Manufacturer Bell & Gossett		
Model/Size 4" BB		
Type Pump Centrifugal		
Impeller Size 9 1/2" dia.		
	Rated	Actual
GPM	480	454
Total Ft. Head	78	82
RPM	1750	1750

MOTOR		
Manufacturer B & G Serial No. F21894		
Frame No. 254T Svc. Factor 1.15		
	Rated	Actual
HP, Nameplate	15	
Amps, L_1 L_2 L_3	42	
Voltage, L_1 L_2 L_3	230	
RPM	1750	
Phase	3	

PUMP PRESSURES		
		Actual
Static Hd (Pump Off)		
Discharge		90 FT
Suction		30 FT
		60 FT
Block Off: (Running, no flow)		
Discharge		150 FT
Suction		45 FT
Total		95 FT
Running:		
Discharge		100 FT
Suction		18 FT
Total		82 FT

STARTER
Manufacturer G. E. Model
Size 2 Class
Overload: Required Size : R 43.1
Actual:

BHP $[HP_{np} \times \frac{A_a}{A_r} \times \frac{V_a}{V_r}]$

$$\text{KWH Per Year} = \frac{\text{Volts} \times \sqrt{3}^{*} \times \text{Avg Amps} \times \text{Yearly Hours of Operation} \times \text{PF}}{1000}$$

= ______ *(3 phase)

Remarks ____________

Wendes Engineering and Contracting Services 7/80 Form TAB 210

Figure 1-12. Sample Filled Out Pump Test Report Form

WATER BALANCE REPORT PRESSURE DIFFERENTIAL

This water balance report is used for readings with fluid flow measuring devices such as differential Pitot tubes, circuit setters, venturis, etc.

The pressure drop across the flow measuring device is measured and correlated to a factory established curve furnished with the measuring device to determine the gpm flow.

The correct units, PSL feet or inches must be used to correlate to factory curve.

Percent of design is shown on the sample report in preliminary readings in the proportionate method of balancing.

Fill out the flow or *pressure drop water balance report.* List primary and secondary circuits in groups and in sequence.

List valves and terminals in the sequence they occur starting at the pump, along with identification and location if needed. Indicate the size. Enter the required cfm and the differential pressure reading required for the particular flow measuring device being used or from the manufacturer's published pressure drop across the item at design flow. List bypasses.

When proportionately balancing terminals, list percents of design in preliminary readings and gpm only for the final.

Flow or Pressure Drop

WATER BALANCE REPORT

Job North High School Job No. C-150 Date Aug. 15, 1981

Location Penthouse System S-1 thru S-5

Instrument Used Annubars Tested by: HW

ΔP: Inches water reading on Annubar

TAG NO.	SIZE	REQUIRED		BY PASS		PRELIMINARY READINGS						FINAL	
		ΔP[1]	GPM	ΔP	GPM	ΔP	%	ΔP	%	ΔP	%	ΔP	GPM
Total Flow Station	6" ϕ	10"	480				99%				94%		454
S-1	2½" ϕ	8.4"	72				120		95%		93		67
S-2	2½" ϕ	6"	60				105		90		97		58
S-3	3" ϕ	8"	108				90		83		94		101
S-4	3" ϕ	8"	108				83		79		95		103
S-5	3½" ϕ	6"	132				73		77		95		125
Total			480										454

1. ΔP: Can represent flow station differential or pressure drop across item.

Remarks

Wendes Engineering and Contracting Services 7/80 Form TAB 211

Figure 1-13. Sample Filled Out Water Balance, Pressure Report Form

WATER BALANCE REPORT THERMAL DIFFERENTIAL

If there is no flow measuring device installed in the systems to measure the flow through a heating or cooling coil, the flow can be approximated indirectly with a thermal method of measurement.

The air entering and leaving temperatures, air flow and water entering and leaving temperatures are measured.

The Btuh of heat transfer is then calculated from the measured air temperatures and cfm. This Btuh is then plugged into the gpm formula along with the fluid temperatures in and out and the gpm is calculated.

Fill out *temperature water balance report* for systems without flow measuring stations at terminals, which are generally reheat coils, induction units and baseboard radiation units. List the coil identification and room number in sequence, from the pump out; list primary circuits and secondary circuits as separate groups. Indicate sizes, design entering and leaving water temperature, and entering and leaving air temperatures.

WATER BALANCE REPORT

Temperature Measurements

Job SEARS Job No. C-120 Date 5-17-88

Location CHICAGO System S-2

Instrument Used ______ Tested by: HCW

COIL	DESIGN						READINGS					CALCULATED	
	GPM	CFM	EWT	LWT	EAT	LAT	EWT	LWT	EAT	LAT	CFM	BTUH	GPM
S-1	36	9,400	45	55	75	57	43	55	74	57	9200	206	34
S-2	34	8,700	45	55	75	57	45	54	75	59	8500	209	32
S-3	50	12,000	45	55	75	57	44	55	73	58	12,100	252	48

Remarks ______

Wendes Engineering and Contracting Services 7/80 Form TAB 212

Figure 1-14. Sample Filled Out Water Balance, Thermal Report Form

CHILLER TEST REPORTS

1. It is vital that the chiller have the correct capacity and be performing as required and produce the proper temperature drop for the chilled water system.

2. Fill out chiller (compressor and condenser), motor and starter descriptive *data* and ratings from equipment drawings and/or blueprints.

3. Check compressor, condenser and motor *nameplates;* make physical inspection for further data and enter.

4. Check refrigerant oil levels.

5. Check starters and overload sizes.

Take electrical readings, chilled and condenser water pressure, temperature and flow readings, and refrigerant pressure and temperature readings.

The Freon pressures and temperatures on the compressor, *evaporator* and *condenser* are normally checked out by the manufacturer in the start-up and are not generally part of the water balance.

Fill in design water pressures, temperatures and flows on the evaporator and condenser.

Fill in design electrical data, hp, kW, voltage, phase, etc., on the compressor and starter.

The conditions at test time will be filled in at the job site; refrigerant and oil levels, water control settings, temperature and pressure cutouts and purge operation.

CHILLER TEST REPORT

Job North High School Job No. C-150 Date Aug. 15, 1981

Location ________ System ________

Equipment Location Mechanical Rm Serves Bldgs. A, B, C Tested by: HW

COMPRESSOR DATA	
Manufacturer	Westinghouse
Model/Size	PE063JAQ/20V/FA2/H02
Type	Package Centrifugal
Capacity 200 tons @ 480 GPM	
Refrigerant 12	Pounds 710
KW 152	KW Per Ton .76
Serial No.	WH 259100

COMPRESSOR MOTOR	Rated	Actual
Manufacturer Westingh.' Serial No.		
Frame No. Type Frame ☐T ☐U		
Svc. Factor: 1.15		
HP, Nameplate	200	
BHP $[HPnp \times \frac{Aa}{Ar} \times \frac{Va}{Vr}]$		
Amps, L_1 L_2 L_3	234	
Voltage, L_1 L_2 L_3	460	
RPM	1800	
Phase	3	

COMPRESSOR	Design	Actual
Suction Pressure		
Suction Temp.		
Discharge Press.		
Discharge Temp.		
Oil Temp/Press.		

STARTER	
Manufacturer Westingh.'	Model
Size 5	Class NEMA 1
Overload: Required Size : SR 240	
Actual:	

EVAPORATOR	Design	Actual
Refrig. Pressure	37 psig	
Refrig. Temp.	40 F	
Ent. Water Pressure		
Lvg. Water Pressure		
Ent. Water Temp.	54 F	
Lvg. Water Temp.	44 F	
Flow GPM		

CONDENSER	Design	Actual
Liquid Line Pressure		
Liquid Line Temp.		
Ent. Water Press.		
Lvg. Water Press.		
Ent. Water Temp.	85 F	
Lvg. Water Temp.	95 F	
Flow GPM		

CONDITIONS	
Refrigerant Level	✓
Oil Level	✓
Percent Cylinders Unloaded	
Chilled Wat. Control Setting	
Condenser Wat. Control Setting	
Low Wat. Cutout Temp. Setting	
Low Pressure Cutout Setting	
High Pressure Cutout Setting	

$$\text{KWH Per Year} = \frac{\text{Volts} \times \sqrt{3} \times \text{Avg Amps} \times \text{Yearly Hours of Operation}}{1000}$$

KW's = ________

Remarks ________

☑ Purge Operation Checked

☑ Crankcase Heater Checked

Wendes Engineering and Contracting Services 7/80 Form TAB 214

Figure 1-15. Sample Filled Chiller Test Report Form

Chapter 2

HVAC Energy Auditing Procedures and Forms

The purpose of an energy audit is to determine the energy consumption and costs of the overall building and of its specific components, the structure, systems and equipment. It is to generate *energy improvement* options, to project *energy savings,* to estimate the *costs* of energy improvements, calculate *paybacks,* and on this basis, evaluate the various options.

A good audit is diagnostic in nature, develops a valid prognosis of the causes of energy wastes, and leads to scientifically established remedies.

Well designed HVAC auditing forms are indispensable to these ends.

HVAC ENERGY AUDITING PROCEDURES

There are two basic phases or types of audits, short walk through audits and in-depth detailed audits, of either the entire building or of only selected parts of a building.

Phase One: Walk Through Energy Audit Procedure

1. Make an initial walk through inspection to become *familiar* with the building, systems, equipment, maintenance, operation status, etc.

2. Study the *plans* and specifications and become familiar with the building, systems, capacities, equipment, etc.

3. Talk briefly with the building operating personnel, owner, occupants, etc. about HVAC systems, comfort, problems, etc.

4. Examine the overall building *energy consumption history* from the owner if available. If not, get complete energy consumption history on gas, oil, electricity, etc., from utility companies and fuel suppliers.

   Compare the *Btu consumption per sq. ft.* per year with similar buildings and determine the degree of variance.

5. List *maintenance,* cleaning, adjustment, repairs and *balancing* needed to this point. Determine what maintenance and repairs must be done before the detailed audit can be performed.

6. Take some *spot test readings* if needed.

7. If a more *extensive audit is* needed, determine what test readings, inspections, analyses, calculations, etc. are required and estimate the time and costs involved.

8. Fill out *Building and Systems Description* report.

9. Write a list of existing energy *problems.*

10. List obvious and *potential energy savings* improvements. Further develop the most promising energy improvements.

11. If the walk through audit is sufficient, calculate energy savings for the various energy improvements, estimate retrofit costs and calculate paybacks.

12. *Select* with owner which energy improvements to proceed with and assign priorities. Properly engineer retrofit work and proceed.

Phase Two: In Depth HVAC Energy Audit Procedure

Field Surveys

1. Make thorough *inspection* of building systems and equipment and become thoroughly familiar with them. Check out operations, performance, maintenance, malfunctions, comfort, problems, etc.

2. Check *nameplate* data on equipment.

3. Conduct in-depth *interviews* with *building personnel.* Review maintenance, scheduling, performance, comfort and problems of building, equipment and systems.

4. Become familiar with actual *hours of operation* of systems and equipment, and the hours of *occupancy* by personnel.

Energy History

5. Study and analyze a 3-year history of the building's electrical and fuel *energy con-*

sumption. Compare with building consumption indexes of similar buildings.

Field Tests

6. Take *test readings* of actual flows, temperatures, pressures, rpm's, amps, volts, etc., at HVAC equipment.

   Check pressure drops across filters, coils, strainers, etc. Check outside air flows at minimum and maximum.

   Monitor readings over a period of time with a recording instrument where required.

   Check lighting levels.

Seasonal and Peak Energy Calculations

7. Determine the actual *seasonal and peak energy consumption* and efficiencies of specific systems and equipment, etc. based on tests and other data.

8. Calculate the peak and seasonal *heating, cooling and cfm loads* actually required to meet current conditions for the overall building and various areas of the building. *Compare* with the design and existing capacities.

Evaluation of Energy Improvements

9. List all *problems* with the building, systems and equipment.

10. Generate *energy improvements* and develop those with the most potential. Write a list of improvements.

11. Calculate the potential *energy savings* in terms of Btu's and kWh's, and in costs.

12. *Estimate costs of retrofit work.*

13. *Calculate paybacks* and returns on investment.

Review and Decisions

14. *Review* with owner or his representatives:
    Problems Costs of Improvements

Energy Improvement Options　Payback Return on Investment
Potential Savings

Consider a *change* only on one portion of the recommended energy improvements to test and validate the savings and to observe the effects.

15. *Select* with the owner which energy improvements to proceed with and assign priorities.

Engineering and Construction

16. *Properly engineer* the owner retrofit work, prepare drawings and write specifications.

17. Obtain quotations, review contracts and *proceed* with the *retrofit work.*

18. *Monitor* units of energy and cost savings after put into operation. Make adjustments and *modifications* as required.

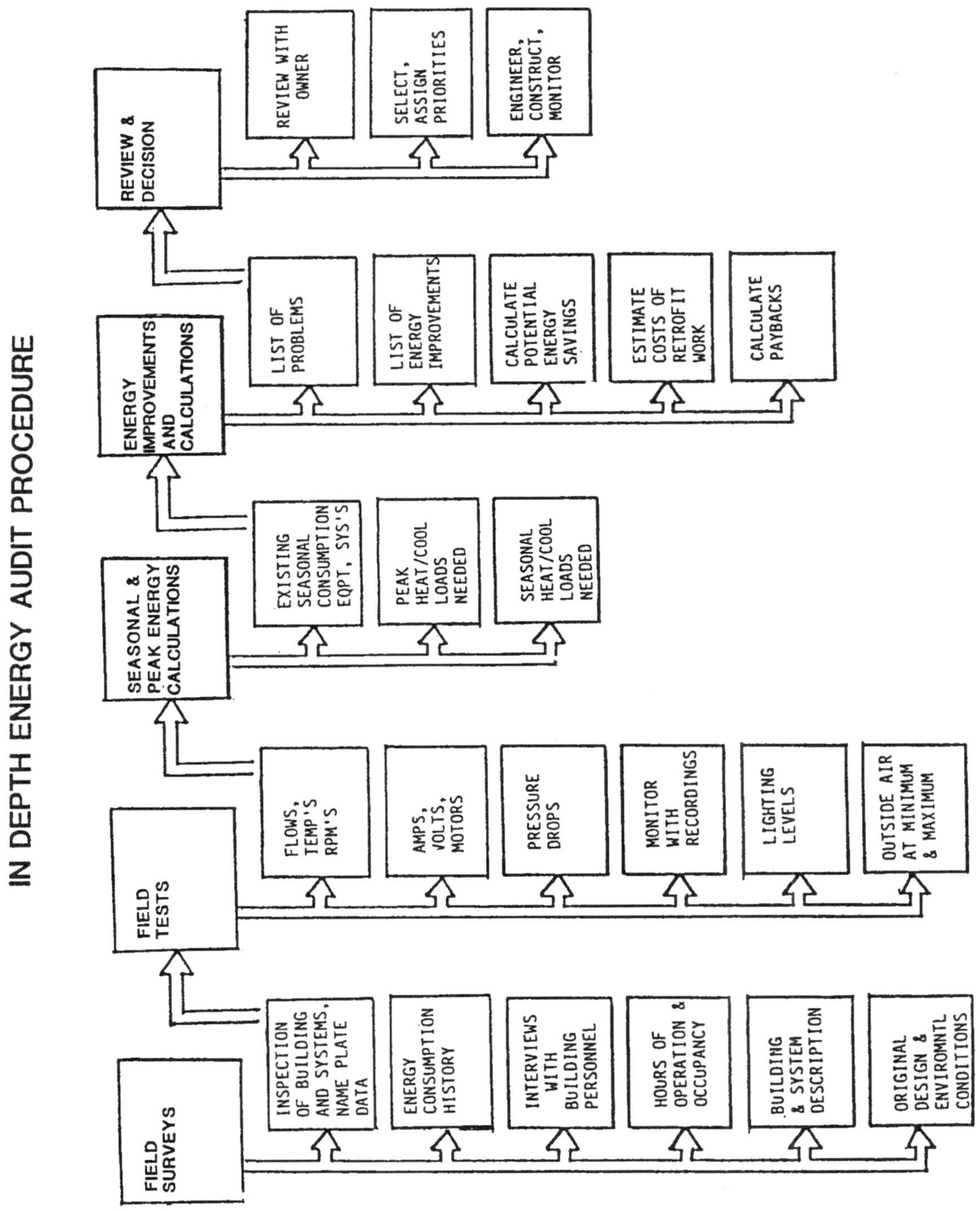

Figure 2-1. In-depth HVAC Audit Procedure Diagram

READINGS REQUIRED FOR THOROUGH HVAC AUDIT OF SYSTEM

Each HVAC system is somewhat unique and its particular characteristics can only be identified by *inspection and measurement.* Information required to understand the present operation of a system and to provide a basis for deciding which modifications are likely to prove beneficial is tabulated below.

ELECTRICAL READINGS	(Amps, Volts, Power Factors, kW) Fans Pumps Compressors Condensers Chillers Lights Owners Operating Equipment
AIR FLOW RATES	Total supply air from fan Total return air to fan Total outdoor air Trunk ducts Terminal units Air cooled condenser
AIR PRESSURE READINGS	Suction and discharge of fans Drops across coils, filters, etc.
WATER FLOW RATES	From pumps Through boilers Through chillers Cooling towers Heat exchangers Coils and terminal units

WATER PRESSURE READINGS	Suction and discharge of pumps Drops across strainers, coils etc. Drops across boilers, chillers, condensers
TEMPERATURES, AIR	Outdoor air db and wb Return air db and wb Mixed air entering coils, db and wb Supply air leaving coils, db and wb Hot deck Cold deck Air at terminals Conditioned areas db and wb
TEMPERATURES, WATER	Boiler supply and return Chiller supply and return Condenser supply and return Coil supply and return Heat exchanger supply and return
REFRIGERANT TEMPERATURES	Hot gas line
OVERALL BUILDING ENERGY READINGS	Suction line At gas meter with all heating on At electric meter with only lights on At electric meter with HVAC units on At electric meter with refrigeration on

Energy conservation must be approached in a systematic manner rather than considering individual items *out of context.* Systems do *not operate in isolation* but depend on and react with other systems. It is important to recognize this interaction of systems as modifications to one will cause a reaction in another which may be either beneficial or counterproductive.

AVERAGE ANNUAL ENERGY PERFORMANCE IN BTU'S PER SQUARE FEET

Heating and Cooling Degree Day Region

Building Type	National	1	2	3	4	5	6	7
Office	84,000	85,000	76,000	65,000	61,000	51,000	50,000	64,000
Elementary	85,000	114,000	70,000	68,000	70,000	53,000	48,000	57,000
Secondary	52,000	77,000	65,000	55,000	51,000	37,000	41,000	34,000
College/Univ.	65,000	67,000	70,000	46,000	59,000			83,000
Hospital	190,000		209,000	171,000	227,000	207,000		197,000
Clinic	69,000	84,000	72,000	71,000	65,000	61,000	59,000	59,000
Assembly	61,000	58,000	76,000	68,000	51,000	44,000	68,000	57,000
Restaurant	159,000	162,000	178,000	186,000	144,000	123,000	137,000	137,000
Mercantile	84,000	99,000	98,000	86,000	81,000	67,000	83,000	80,000
Warehouse	65,000	75,000	82,000	65,000	50,000	38,000	37,000	39,000
Residential	95,000	99,000	84,000	94,000	125,000	90,000	93,000	106,000
Non-Housekeeping								
High Rise Apt.	49,000	53,000	53,000	52,000	53,000	84,000	20,000	

February 1978, HUD-PDR-290

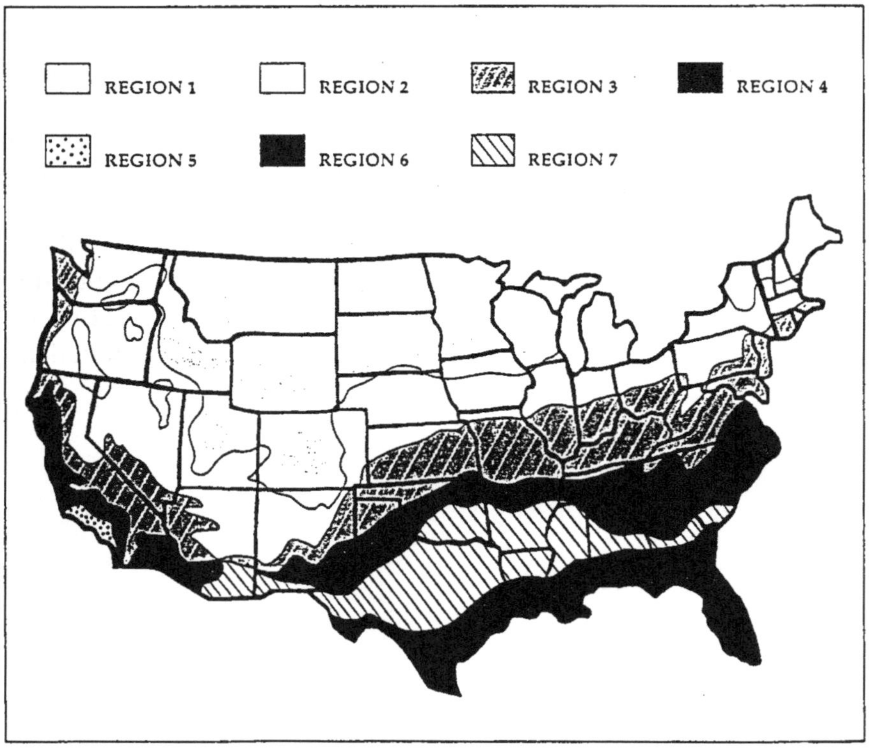

Figure 2-2

SAMPLE AUDIT AND FORMS

SAMPLE HVAC ENERGY AUDIT OF A SUBURBAN OFFICE BUILDING

This 90,000-sq.-ft suburban office building was built around 1967 before the energy crisis occurred and it incorporates in its design and operation the great energy waste of that era. Overall, the FWAC, electrical and plumbing systems in the building consumed *276,000 Btu per sq. ft.* per year for a total consumption of *24.9 bill Btu* for 1994. The energy costs were *$3.80 per sq. ft.* and totaled *$339,800* for the year.

The *targeted* energy reduction is 50 percent, reducing the Btu per sq. ft. from 276,000 to 138,000 Btu per sq. ft. with savings of about $1.90 per sq. ft, or $171,000 per year.

PROBLEMS WITH EXISTING HVAC SYSTEMS AND BUILDINGS

1. The building has two energy-wasting HVAC systems which simultaneously heat and cool, an interior high pressure *terminal reheat* system and a high pressure perimeter *induction system* with about one third primary air taken from the outside.
2. The 310-ton *chiller,* 6,000,000 Btuh boiler and the hot- and chilled-water pumps must run year round because of the terminal reheat system, wasting a great deal of energy.
3. The *computer room* operates 24 hours a day, weekdays, with a constant cooling load demand forcing the chiller to run year round and sporadically at nights.
4. The *chiller* and *boiler* are *oversized;* they cycle and run at inefficient levels.
5. *Excessive minimum outside air is* brought into the building. The settings of the dampers are off and they leak.
6. The *lighting levels* are excessive.
7. *Thermostats* are set too high or low in the winter and too low in the summer.
8. The oil-fired *boiler* is inefficient, scaly, with a poor combustion efficiency of 60 percent.
9. *Maintenance* is poor. Filters, coils and strainers are generally dirty. Many control valves, automatic dampers and thermostats are out of calibration, malfunctioning or miss-set.
10. Systems are *out of balance.*
11. *Fans* are oversized for load, pumping out more air than required, running on demand 24 hours a day year round.
12. *Pumps* are oversized for load, pumping out more gpm's than required, running on demand 24 hours a day year round.
13. *Starting and stopping times* of HVAC equipment not optimized.
14. Paying higher *demand* rates than need be.
15. *Power factors:* on underload motors not controlled.

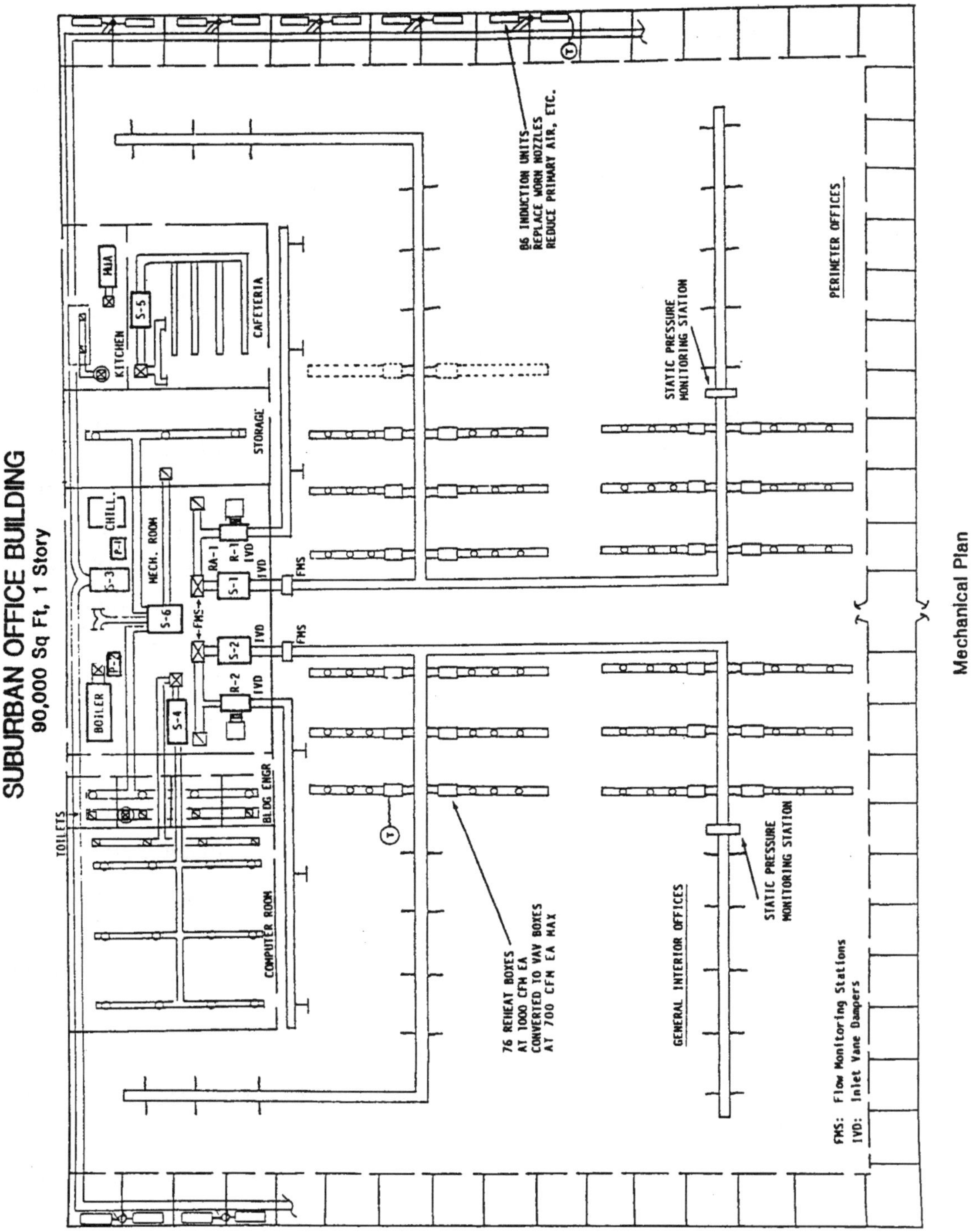

Figure 2-3. Suburban Office Building HVAC Diagram

HVAC ENERGY AUDITING REPORT

Date ____________________

Job ____

Location ____

Architect ______________________ Phone ______________________

Engineer ______________________ Phone ______________________

Auditing Contractor ______________________________________

Address ____

City __________________________ State ________ Zip ____________

Phone ______________________ Fax Number ______________________

Figure 2-4 Cover Sheet for Energy Auditing Report

BUILDING AND SYSTEM DESCRIPTION FORMS

The Building and System Description information sheets are extremely important forms to be thoroughly filled out for a complete, effective and worthwhile energy audit to be made, and for a worthwhile energy conservation program to be brought about for the building.

The information entered in this initial form is very critical. It involves a thorough study of the original plans and specifications on the building, a preliminary job site survey, and gathering and evaluating energy utility bills and consumption for the building. The areas of coverage for the forms is as follows:

1. The basic geographical data at the start, as well as the name of the project.
2. The description of the building such as size, occupants, type spaces etc.
3. More specific construction details as to windows, walls, floors, roof and so on.
4. Hours of occupancy.
5. Heating and cooling system description.
6. Actual existing annual energy consumption of the building for heating, cooling, electrical power and lighting, for evaluation and comparison later.
7. The original environmental design conditions of heating, cooling and air flows.
8. Specifications on original or actual lighting levels.
9. Type of electrical service.
10. Connected electrical loads in kW's.

BUILDING AND SYSTEM DESCRIPTION

Name Suburban Office Building

Location

Latitude 41°N Elevation 658 Ft When Built 1967

A. CATEGORY OF STRUCTURE

Office Building

B. BUILDING DESCRIPTION

Area, Sq Ft: 90,000 Number of Floors: 1

Volume, Cu Ft: 1,260,000

Number of Occupants: 400 Sq Ft/Person: 225

Types of Areas: Offices, computer room, kitchen, dining room, employee lounges, storage, mechanical room

C. CONSTRUCTION DETAILS

Glass: Single pane, clear U=1.13, no shading, no drapes or blinds, sealed, aluminum frame

Exterior Walls: 8" brick and block, lathe and plaster, R-6, U factor .167

Roof and Ceilings: Built up tar and gravel on 2" rigid insulation & metal deck, suspended accoustical ceiling. U factor .11

Floors: Concrete slab, 2 BTUH per sq ft

Total Exposed Wall Area Sq Ft: 10,430

Total Glass Area Sq Ft: 6,650 Percent 39

D. HOURS OF OCCUPANCY AND OPERATION

Working Hours: 8am to 6pm weekdays, 9am to noon sat.

Lighting Hours: 8am to 9pm weekdays, 3 hrs sat.

HVAC Hours: Cooling and heating year around

Janitorial Cleanup Times: 6pm to 9pm weekdays

Computer Room: 24 hrs per day, 365 days per year

Other:

E. HEATING AND COOLING SYSTEMS DESCRIPTION

Interior office spaces, high pressure terminal reheat system

High pressure perimeter induction system, 100% primary air

Chilled water cooling: 1 centrifugal chiller

Hot water heating, oil fired

Kitchen, Dining: Single zone low pressure

Computer Room: Single zone low pressure

Wendes Mechanical Consulting Services Co. Form AUD 301

Figure 2-5. Sample Filled Out Building and System Description Form

F. ANNUAL ENERGY CONSUMPTION

Total Heating, Cooling, Electrical, Lighting Per Yr:

Total BTU: 24.9 Bill. BTU Per Sq Ft: 276,600

Total Energy Costs: $339,862 Costs Per Sq Ft: $3.78

Electrical, Total KWH: 3,829,411 KWH/Sq Ft: 42.55

Total Elec. Costs: $268,062 Costs/Sq Ft: $2.98

Heating Fuels, BTU Per Yr: 11.8 Bill. Per Sq Ft: 131,110

Total Fuel Costs: $71,800 Costs/Sq Ft: $.81

G. ORIGINAL ENVIROMENTAL DESIGN CONDITIONS

Heating

Peak Heat Loss BTUH: 4.8 Mill. Output Degree Days: 6,000

Design Temperatures: -10F, 74F

Avg Winter Temp.: 35 Avg Winter Hours: 4,800

Cooling

Peak Heat Gain BTUH: 310 tons Degree Days: 682

Design Temperatures: 94F DBT, 75F WBT Outdoors
74F DBT 50% RH Indoors

Avg Summer Temp.: 75 Avg Summer Cool Hours: 900

Air and Hydronic Flows

Supply CFM: 121,000 CFM/Sq Ft: 1.34

Exhaust Air CFM: 7,300 Exh Air/Sq Ft: .08

Min Outside Air CFM: 30,000 OA Per Person, Sq Ft: 75 or .3/sq ft

Make Up Air CFM: 5,400

HVAC GPM: 1,954 Domestic GPM:

H. LIGHTING

Levels in Foot Candles: 100-200

Levels in Watts/Sq Ft: 4.0 avg.

Type: Fluorescent

I. ELECTRICAL SERVICE

Type: Underground Metering: Primary

Voltage: 277/480V, 3 phase, 4 wire, wye

J. CONNECTED ELECTRICAL LOADS (KW's)

Lighting: 360 KW Office Equipment: 37+10 comp KW

Heating and Cooling Equipment: 286 KW

Air Handling and Exhausts: 160 KW

Cooking: 50 KW Machinery:

Total: 903 KW

Wendes Mechanical Consulting Services Co.

Figure 2-5. Sample Filled Out Building and System Description Form (*Cont'd*)

(Proceed to next page...)

ELECTRICAL CONSUMPTION HISTORY PER MONTH

The electrical consumption history per month form is essential for transferring the electric consumption kWh's, demand charges and costs from the electrical utility bills to a list that can be added up for yearly totals, and studied. Then ratios such as cost per square foot of building can be calculated and noted at the bottom—for evaluation against other building costs, budget figures, benchmarks etc.

ELECTRICAL CONSUMPTION HISTORY PER MONTH

BUILDING SUBURBAN OFFICE BUILDING YEAR 1984

SIZE SQ FT 90,000

ELECTRICAL COSTS

MONTH	NO OF DAYS	KWH USED	COST PER KWH	DEMAND PEAK	DEMAND CHARGE	POWER FACTOR ADJ	FUEL ADJ	TOTAL COST
JAN		262,651	0.0670		$769		0.200	$17,598
FEB		285,739	0.0600		659		0.200	$17,144
MAR		219,792	0.0722		769		0.417	$15,875
APR		230,782	0.0670		659		0.417	$15,462
MAY		311,006	0.0760		1,429		0.200	$23,636
JUNE		362,657	0.0780		1,429		0.200	$28,287
JULY		403,318	0.0720		1,319		0.200	$29,039
AUG		429,693	0.0720		1,429		0.200	$30,938
SEPT		422,001	0.0750		1,539		0.200	$31,650
OCT		350,568	0.0690		1,319		0.200	$24,189
NOV		275,839	0.0620		769		0.200	$17,102
DEC		270,161	0.0640		659		0.200	$17,290
TOTAL		3,824,207	0.0701	0	12,748			$268,210
AVG/MO		318,684	0.0701	0	1,062			$22,351

1. Average KWH per sq ft of building per year................. 42.49
   Average electrical cost per sq ft of building per year..... $2.98
2. BTUH equivalent for year, (KWH=3413 BTUH)................ 13.05 billion
   BTUH average per month.................................... 1.09 billion
3. Average cost per million BTUH............................. $20.55
   Average BTU per sq ft of building per year for electrical.. 13,052
4. Average KWH per hour...................................... 437

Wendes Mechanical Consulting Services Co. Form AUD 303

Figure 2-6. Sample Filled Out Electrical Consumption History Per Month Form

HVAC EQUIPMENT SCHEDULES

Overall listings of all the HVAC equipment, air handling units, fans, pumps, and chillers are required for a concrete comprehensive view of the systems in the building and easy review of their performance.

It is necessary to know the key design figures of the major HVAC equipment as indicated on the plans, and to have specifications of flow, pressures, temperatures, electrical, etc. Then take actual readings to know what the current figures are.

From these two sets of figures discrepancies between design and current actuals are noted, evaluated, and the systems corrected as required. Then actual figures are rechecked and become the basis for determining the actual energy saved after the retrofit and for monitoring purposes.

Separate reports for taking readings and for recording in more detailed for major equipment follow in forthcoming pages.

HVAC EQUIPMENT SCHEDULE

Job: Corp. Office Bldg. Location: Date: 11-8-84

SYSTEM	SERVES	EQPT LOCAT	TYPE EQPT	☒ FAN ☒ CFM ☐ GPM	☐ PUMP RPM	☒ SP ☐ FT HD	OUTLET VELOC	MOTOR HP	AMPS	VOLTS	COOLING COIL MBH	TONS	GPM	HEATING COIL MBH	GPM
S-1	Interior Office East		AHU	38,000	1585	8"		75	96	460	1140	95	228	800	80
S-2	Interior Office West		AHU	38,000	1585	8"		75	96	460	1140	95	228	800	80
S-3	Perimeter Offices		AHU	7,000	2832	7"		10	28	230	600	50	120	760	76
S-4	Computer Rm.		AHU	10,400		2"		10	28	230	312	26	62	125	13
S-5	Cafeteria & Kitchen		AHU	9,800		1½"		5	15	230	300	25	48	70	70
S-6	Toils, Stor, Mech. Rm.		MZ	7,200		1½"		5	15	230	216	18	43	89	89
	TOTALS			121,200				180			3708	303	729	2644	408
MUA-1	Kitchen		MUA	5,400	760	1"		3	9	230	---	---	---	467	---
	Motors:		3 P.H.,	1 Cycle,	1750 RPM										

FAN EQUIPMENT SCHEDULE

Job: Corp. Office Bldg Location: Date: 11-8-84

SYSTEM	SERVES	EQPT LOCAT	TYPE EQPT FANS	☒ FAN ☒ CFM ☐ GPM	☐ PUMP RPM	☒ SP ☐ FT HD	OUTLET VELOC	MOTOR HP	AMPS	VOLTS	COOLING COIL MBH	TONS	GPM	HEATING COIL MBH	GPM
R-1	Interior Office East		Centrif	32,000	509	2"		15	42	230					
R-2	Interior Office West		Centrif	32,000	509	2"		15	42	230					
TE	Toilets		Pre	675	700	3/8"		1/4	3	115					
KE	Kitchen Hood		Pre	5,400	760	1"		3	9	230					
E-1	Conference Rm		Wall	600	960	1/4"		1/4	3	115					
E-2	Conference Rm		Wall	600	960	1/4"		1/4	3	115					
	TOTALS			71,275				34							

PUMP EQUIPMENT SCHEDULE

Job: Corp. Office Bldg Location: Date: 11-8-84

SYSTEM	SERVES	EQPT LOCAT	TYPE EQPT	☐ FAN ☐ CFM ☒ GPM	☒ PUMP RPM	☐ SP ☒ FT HD	OUTLET VELOC	MOTOR HP	AMPS	VOLTS	COOLING COIL MBH	TONS	GPM	HEATING COIL MBH	GPM
P-1	Chilled Water		Pump	744	1750	93		30	40	460					
P-2	Hot Water		Pump	466	1750	80		20	27	460					
P-3	Cooling Tower		Pump	744	1750	93		30	40	460					
CH-1	Chiller							300 (224w)	339	460		310	744		

Figure 2-7. Sample Filled Out HVAC, Fan and Pump Equipment Schedule

ELECTRICAL CONSUMPTION GRAPHS

The electrical consumption figures can be punched into a spreadsheet and graphed automatically as shown.

This helps greatly to visualize and analyze the electrical energy situation in terms of overall costs, demand charges and kWh per month. Peaks and valleys are more easily recognized and dealt with.

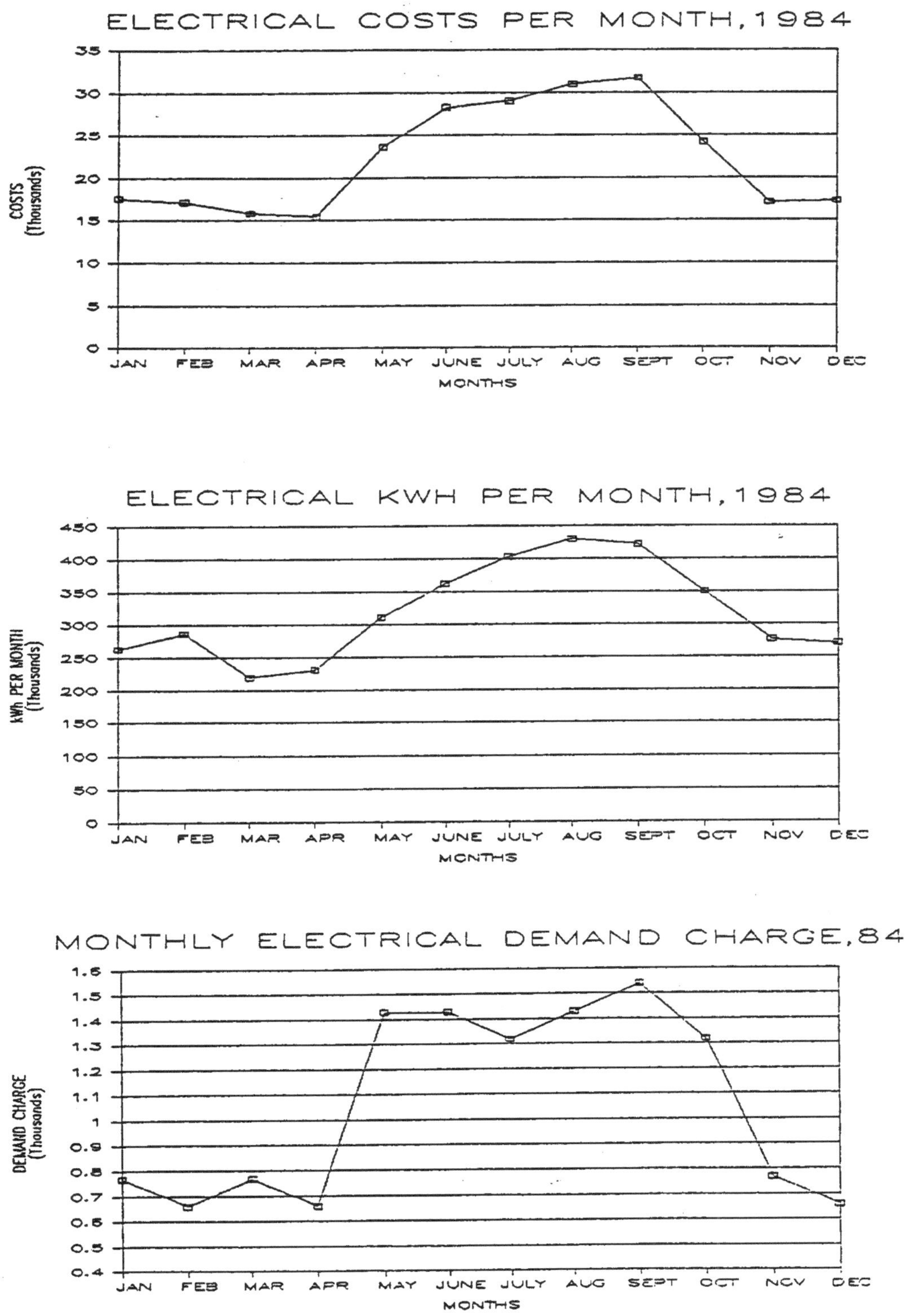

Figure 2-8. Electrical Consumption Graph Samples

EQUIPMENT TEST REPORTS

Detailed test reports for each major piece of equipment in the building involved with the energy audit and retrofit are very important for accuracy, auditing, redesigning and monitoring later.

These reports record the key performance figures such as:

- gpm
- cfm
- Amp draws
- Pressures
- Temperatures, etc.

They also record important information such as:

- Detailed description of the equipment
- Manufacturers
- Model numbers.
- Maximum amp draws on motors
- Service factors
- Settings of dampers and valves
- Pressure drops

Test reports are required on major equipment such as fans, pumps, chillers, condensers, boilers, etc.

The following three equipment test reports are also found in the testing and balancing chapter with further information on them.

FAN TEST REPORT

Job North High School Job No C-150 Date Aug. 1, 1981
Location ______ System S-2
Equipment Location Mezzanine Serves Lunchroom Tested By: HW

☑ Air Handling Unit ☐ Roof Top Unit ☐ Furnace ☐ Supply Fan ☐ Exhaust Fan ☐ Pkg Unit
☑ LP ☐ MP ☐ HP ☑ Constant Volume ☐ VAV

FAN DATA	
Manufacturer	Barry
Model Size	AF 7245 DWDI
Type Fan	☑ Centrigal ☐ Roof Exhaust ☐ Inline ☐ Vane Axial ☐ Prop.
Type Wheel	☐ Backward Incline ☑ Air Foil ☐ Forward Curve ☐ Paddle Wheel
Wheel:	☑ Alignment OK ☑ Gap ☑ Fastened ☑ Clean
Belts (2) B131	C to C Distance 52"
Pulleys: Fan Dia. 10"	Mot. Dia. 5"
Motor Movement 2" ±	Belts
Bearings ✓	☑ Cut Off Plate OK

MOTOR		
Manufacturer GE	Serial No.	
Frame No. 184T	Type Frame ☐ T ☑ U	
Service Factor: 1.15	Rated	Actual
HP, Nameplate	5	5
BHP $[HPnp \times \frac{Aa}{Ar} \times \frac{Va}{Vr}]$	3.53	3.84
Amps, L_1 L_2 L_3	15.2	12.2
Voltage, L_1 L_2 L_3	230V	220V
RPM	1750	1750
Phase	3	3

FAN PERFORMANCE		
	Design	Actual
Fan CFM	9,800	10,160/104%
Outlet CFM Total	9,800	9,756/99%
Fan RRM	985	992
Fan S.P.	1½"	1.6"

STARTER	
Manufacturer GE	Model 141R
Starter Size 0	Class I
Overload: Required Size	CR 15.4
Actual	CR 15.4

STATIC PRESSURE DROPS	Upstream	Downstream	Total Drop
Filter	.2"	.4"	.2"
Heat. Coil	.4	.6	.2
Cool. Coil	.6	1.0	.4
Fan Inlet			1.0
Fan Discharge			.6
Total Fan S.P.			1.6"

W.G.

CONDITIONS			
Vortex Damper Position	—		
Outside Air Damper Setting	4000 CFM		
Return Air Damper Setting	6160 CFM		
Filter Conditions	Clean		
Coil Conditions	Clean		
Temperatures			
OA: @ 40%	40 F DB	WB	RH
RA:	70 F DB	WB	RH
Mixed Air:	DB	WB	RH
Discharge	DB	WB	RH
Space:	DB	WB	RH
Duct Temp. Drop	DB		

Remarks ______

PROBLEMS:
☐ Too much air ☐ SP Low ☐ Too Hot
☐ Too little air ☐ SP High ☐ Too Cold
☐ Air Noises ☐ Fan Noises
☐ Oversized equipment ☐ Undersized Equipment
☐ Other ______

Wendes Engineering and Contracting Services 7/80 Form TAB 201

Figure 2-9. Sample Filled Out Fan Test Report

PUMP TEST REPORT

Job North High School Job No. C-150 Date Aug. 15, 1983

Location ____________ System ____________

Equipment Location Penthouse Serves Bldgs A, B, C Tested by: HW

PUMP DATA		
Manufacturer Bell & Gossett		
Model/Size 4" BB		
Type Pump Centrifugal		
Impeller Size 9 1/2" dia.		
	Rated	Actual
GPM	480	454
Total Ft.Head	78	82
RPM	1750	1750

MOTOR		
Manufacturer B & G	Serial No. F21894	
Frame No. 254T	Svc. Factor 1.15	
	Rated	Actual
HP, Nameplate	15	
Amps, L_1 L_2 L_3	42	
Voltage, L_1 L_2 L_3	230	
RPM	1750	
Phase	3	

PUMP PRESSURES		
		Actual
Static Hd(Pump Off)		
Discharge		90 FT
Suction		30 FT
		60 FT
Block Off: (Running, no flow)		
Discharge		150 FT
Suction		45 FT
Total		95 FT
Running:		
Discharge		100 FT
Suction		18 FT
Total		82 FT

STARTER	
Manufacturer G. E.	Model
Size 2	Class
Overload: Required Size : R 43.1	
Actual:	

$$\text{BHP } \left[\text{HPnp} \times \frac{Aa}{Ar} \times \frac{Va}{Vr}\right]$$

$$\text{KWH Per Year} = \frac{\text{Volts} \times \sqrt{3}^{*} \times \text{Avg Amps} \times \text{Yearly Hours of Operation} \times \text{PF}}{1000}$$

= [ ] *(3 phase)

Remarks ____________

Wendes Engineering and Contracting Services 7/80 Form TAB 210

Figure 2-10. Sample Filled Out Pump Report

CHILLER TEST REPORT

Job North High School Job No. C-150 Date Aug. 15, 1981

Location ______ System ______

Equipment Location Mechanical Rm Serves Bldgs. A, B, C Tested by: HW

COMPRESSOR DATA	
Manufacturer Westinghouse	
Model/Size PE063JAQ/20V/FA2/H02	
Type Package Centrifugal	
Capacity 200	tons @ 480 GPM
Refrigerant 12	Pounds 710
KW 152	KW Per Ton .76
Serial No. WH 259100	

COMPRESSOR MOTOR		
Manufacturer Westingh. Serial No.		
Frame No. Type Frame ☐T ☐U		
Svc. Factor: 1.15	Rated	Actual
HP, Nameplate	200	
BHP [HPnp × $\frac{Aa}{Ar}$ × $\frac{Va}{Vr}$]		
Amps, L_1 L_2 L_3	234	
Voltage, L_1 L_2 L_3	460	
RPM	1800	
Phase	3	

COMPRESSOR	Design	Actual
Suction Pressure		
Suction Temp.		
Discharge Press.		
Discharge Temp.		
Oil Temp/Press.		

STARTER	
Manufacturer Westingh.	Model
Size 5	Class NEMA I
Overload: Required Size : SR240	
Actual:	

EVAPORATOR	Design	Actual
Refrig. Pressure	37psig	
Refrig. Temp.	40F	
Ent. Water Pressure		
Lvg. Water Pressure		
Ent. Water Temp.	54F	
Lvg. Water Temp.	44F	
Flow GPM		

CONDENSER	Design	Actual
Liquid Line Pressure		
Liquid Line Temp.		
Ent. Water Press.		
Lvg. Water Press.		
Ent. Water Temp.	85F	
Lvg. Water Temp.	95F	
Flow GPM		

CONDITIONS	
Refrigerant Level	✓
Oil Level	✓
Percent Cylinders Unloaded	
Chilled Wat. Control Setting	
Condenser Wat. Control Setting	
Low Wat. Cutout Temp. Setting	
Low Pressure Cutout Setting	
High Pressure Cutout Setting	

$$\text{KWH Per Year} = \frac{\text{Volts} \times \sqrt{3} \times \text{Avg Amps} \times \text{Yearly Hours of Operation}}{1000}$$

KW's = ______

Remarks ______

☑ Purge Operation Checked

☑ Crankcase Heater Checked

Wendes Engineering and Contracting Services

7/80 Form TAB 214

Figure 2-11. Sample Filled Out Chiller Test Report

ELECTRICAL LOADS PER SYSTEM

This valuable form lists the electrical loads and costs separately for each system in the building and allows you to summarize the total for all of the figures and provides an easy review of the systems in recap form.

Data entered cover electrical ratings of the motors:

- Hours of operation
- Actual average electrical loads
- kWh consumed per year
- Electrical utility cost for each piece of equipment for the year.

ELECTRICAL LOADS PER SYSTEM

☒ EXISTING ☐ NEW

BUILDING SUBURBAN OFFICE BUILDING Date 09-27-1985

AIR HANDLING EQUIPMENT ☒ Seasonal ☐ Summer ☐ Winter ☒ Year Around

ITEM	HP OR KWH	RATED VOLTS	RATED AMPS	RATED PH	HOURS OF OPERATION PER YEAR	ACTUAL AVERAGE VOLT LOAD	ACTUAL AVERAGE AMP LOAD	KWH PER YEAR	COSTS PER YEAR
S-1, AHU, INT. EAST	75	460	96	3	5,260	460	77	322,315	$22,562
S-2, AHU, INT. WEST	75	460	96	3	5,260	460	77	322,315	$22,562
S-3, AHU, PERIMETER	10	460	28	3	5,260	460	25	104,648	$7,325
S-4, AHU, COMPUTER RM	10	460	28	3	8,760	460	25	174,280	$12,200
S-5, AHU, CAFETERIA	5	460	5	3	2,500	460	4	7,958	$557
S-6,MZU,TOILETS,STORAGE	5	460	5	3	8,760	460	4.2	29,279	$2,050
MUA-1, KITCHEN	3	230	9	3	2,500	230	7	6,963	$487
R-1, CENTRIFUGAL FAN	15	230	42	3	5,260	230	35	73,253	$5,128
R-2, CENTRIFUGAL FAN	15	230	42	3	5,260	230	34	71,160	$4,981
PRE. TOILET EXHAUST	0.25	115	3	1	8,750	115	2	3,486	$244
PRE, KITCHEN EXHAUST	3	230	9	3	2,600	230	8	8,276	$579
E-1, WALL FAN, CONF.	0.25	115	3	1	2,600	115	2.5	1,293	$91
E-2, WALL FAN, CONF.	0.25	115	3	1	2,600	115	2.5	1,293	$91
(S-1,S-2,S-3,R-1,R-2 off nights and weekends,April thru Nov.)									
TOTAL	217		369				303.2	1,126,520	$78,856

$$\text{KWH/YEAR} = \frac{1.73 \times I \times E \times \text{HOURS}}{1000}$$

COSTS PER KWH = $0.070

Wendes Mechanical Consulting Services Co. Form AUD 305

Figure 2-12. Sample Filled Out Electrical Loads Per System Form

RECAP OF ALL ELECTRICAL LOADS

This form summarizes all the electrical loads in the building or in the complex, from all sources, not just major HVAC equipment.

It includes the total loads from:

- Lighting
- Air handling equipment
- Hydronic heating
- Air conditioning equipment
- Computers
- Office equipment
- Kitchen equipment
- Machinery

The totals from this sheet get transferred to the building and system description form.

ELECTRICAL LOADS PROJECT RECAP

☒ EXISTING ☐ NEW

BUILDING SUBURBAN OFFICE BUILDING Date 11-18-1985

☒ Year Around ☐ Summer ☐ Winter ☐ Peak

ITEM	CONNECTED LOAD KWH	CONNECTED LOAD HP	PH	HOURS OF OPERATION PER YEAR	ACTUAL LOAD KWH	ACTUAL LOAD BHP	ACTUAL KWH PER YEAR	COSTS PER YEAR	PERCENT OF TOTAL
LIGHTING, WINTER	360			2,200	360		792,000	$55,440	0.21
SUMMER	360			1,800	360		648,000	$45,360	0.17
AIR HANDLING EQPT.	160	216					1,126,520	$78,856	0.29
CHILLERS,PUMPS,COOL.TWRS	286	384					980,625	$68,644	0.26
COMPUTERS	10			8,760			87,600	$6,132	0.02
OFFICE EQUIPMENT	37	25		2,600			65,000	$4,550	0.02
KITCHEN EQUIPMENT	50			2,500			125,000	$8,750	0.03
3,829,411 KWH / 90,000 sq ft			=	42.55 KW / sq ft / yr					
$268,062 / 90,000 sq ft			=	$2.98 / sq ft / yr					
3,829,411 KWH x 3416 BTU / KW			=	13.1 Billion equivalent BTU					
TOTAL	903	625					3,824,745	$257,732	1.00

1 HP = 746 WATTS COSTS PER KWH = $0.070

Wendes Mechanical Consulting Services Co. Form AUD 306

Figure 2-13. Sample Filled Out Recap of All Electrical Loads Form

HEAT LOSS CALCULATION FORMS

The original heat loss calculations and capacity HVAC equipment chosen at the time, may have greatly changed over the years and new heat loss calculations are required based on current conditions, to end up with a reliable energy retrofit program.

Lighting loads may have been reduced, insulation added, thermal windows added, temperature settings changed, etc., since the building was built, and the old figures are no longer valid.

New calculations based on the current conditions are required for a valid audit and energy conservation program.

The heading on the heat loss calculation sheet describes the building, its size, design temperatures, etc.

The lines below list all the components involved in the heat loss calculations: roofs, floors, walls, windows, etc., along with dimensions, sq. footage, U factors and temperature differences.

The next column to the right is for the extended total Btu for peak load per hour calculations, and for equipment sizing purposes.

The far right column lists the seasonal hours involved.

Total peak Btu's and tonnage are summed at the bottom.

HEAT LOSS CALCULATION

☐ PEAK PER HR ☒ SEASONAL ☒ EXISTING ☐ NEW

Building SUBURBAN OFFICE BUILDING Date 11-18-1985

Location Latitude 41

Type Building OFFICE AND LABS Stories 1 When Built 1967

Sq Ft Area 90,000 Cubic Ft of Space 1,260,000

X Calculation for Whole Building For Partial Area

Budget Load: Sq Ft 90,000 x BTUYR/Sq Ft 78,500 = 7,065 Mill BTUYR

Outside Design: DB -10 WB Avg OA Temp. 35 Winter Hours 4,800

Inside Design: DB(day) 74 RH DB(night)

ITEM	DIMENSIONS	SQ FT	U	TEMP DIFF	BTU: Per Hour Seasonal	SEASONAL HOURS
ROOF OR CEILING	360 X 250	90,000	0.110	39.00	1,853,280,000	4,800
FLOOR	360 X 250	90,000	0.100	16.00	691,200,000	4,800
GLASS	950 X 7	6,650	1.100	39.00	1,369,368,000	4,800
DOORS						
WALLS	1220 X 14	10,430	0.330	39.00	644,323,680	4,800
COLD INSIDE WALLS						
VENTILATION	1.43 AC/HR CFM=	30,000	1.080	39.00	6,065,280,000	4,800
DUCT LOSSES						
GROSS TOTAL HEAT LOSS					10,623,451,680	
HEAT GAINS; LITES	4W/SQ FT	90,000	4.000	3.416	3,074,400,000	2,500
PEOPLE	No. People =	400	250		200,000,000	2,000
OFF. EQUIPMENT	HP =	25	2,550		127,500,000	2,000
COMPUTERS	KW =	5		3416.00	149,620,800	8,760
TOTAL INTERNAL GAINS					3,551,520,800	
NET TOTAL BUILDING HEAT LOSS					7,071,930,880	78,577*
INPUT TO HEATING EQUIPMENT, efficiency =			.60		11,800,000,000	131,110*

BTUH = Sq Ft x U x Temp. Diff. BTU/YEAR = Sq Ft x U x Avg Temp. Diff. x Winter Hours

* BTU/SQ FT/YEAR

Remarks ____________________

Wendes Mechanical Consulting Services Co. Form AUD 307

Figure 2-14. Sample Filled Out Heat Loss Calculation Form

COOLING LOAD CALCULATION FORMS

The original cooling load calculations and capacity HVAC equipment chosen at the time, may have greatly changed over the years and new cooling load calculations are required based on current conditions, to end up with a reliable energy retrofit program.

Lighting loads may have been reduced, insulation added, thermal windows added, temperature settings changed, etc., since the building was built, and the old figures are no longer valid.

New calculations based on the current conditions are required.

The heading on the cooling load calculation sheet describes the building, its size, design temperatures etc.

The lines below list all the components involved in the cooling loads calculations roofs, floors, walls, windows, etc., along with dimensions, sq. footage, U factors and temperature differences.

The next column to the right is for the extended total Btuh for peak load per hour calculations and for equipment sizing purposes.

Total peak Btu's and tonnage are summed at the bottom.

COOLING LOAD CALCULATION

☒ PEAK PER HR ☐ SEASONAL ☒ EXISTING ☐ NEW

Building SUBURBAN OFFICE BUILDING Date 11-18-1985

Location Latitude 41 Peak Load, HR, MO:

Type Building OFFICE AND LABS Stories 1 When Built 1967

Sq Ft Area 90,000 Cubic Ft of Space 1,260,000

X Calculation for Whole Building For Partial Area

Building Load: Sq Ft 90,000 x Sq Ft/Ton 270 = 333 Tons

Outside Design: DB 95 WB 75 Avg OA Temp. 35 Summer Hours 900

Inside Design: DB(day) 74 RH DB(night) 74

ITEM (Orient.)	DIMENSIONS	SQ FT	U or FACTOR*	TEMP DIFF	SENSIBLE BTUH	TONS	LATENT BTUH
ROOF OR CEILING		90,000	0.110	20.00	198,000	16.50	
GLASS EAST		1,500		15.00	22,500	1.88	
(Solar) SOUTH		2,160		50.00	108,000	9.00	
WEST		1,500		15.00	22,500	1.88	
GLASS SOUTH		5,160	1.130	20.00	116,616	9.72	
(Conduction)						0.00	
WALLS E,W,N		9,040	0.167	20.00	30,194	2.52	
(Conduction)		2,880	0.167	50.00	24,048	2.00	
LIGHTING		90,000	4	3.416	1,229,760	102.48	
PEOPLE, NO.	(NO.)	400	250	250	100,000	8.33	100,000
OFF. EQUIPMENT	MOTORS (HP)	25	2,550		63,750	5.31	
COMPUTERS	(KW)	10	3,416		34,160	2.85	
KITCHEN						0.00	
VENTIL. AIR, Sens	1.43 AC/HR	30,000 CFM	1.08	20.00	648,000	54.00	
VENTIL. AIR, Lat.	1.43 AC/HR	30,000 CFM	0.010	4,840		0.00	1,452,000
TOTAL COOLING LOAD,			SENSIBLE		2,597,528	216	
			LATENT		1,552,000	121	
			GRAND TOTAL		4,149,528	337	267*

BTUH = Sq Ft x U x Temp. Diff.
* U, CLTD, CLF, KW, HP, WD
CLTD = (Temp diff) - (Daily Range + 14)/2
BTU/YEAR = Sq Ft x U x Avg Temp. Diff. x Hrs

* Sq Ft/Ton

Figure 2-15. Sample Filled Out Cooling Load Calculation Form

PEAK HEATING, COOLING AND CFM PER AREA

This very valuable form shows the peak heating and cooling loads for the main areas in the building and the amount of cfm in each area.

The form can be used to examine the original design or what the current loads actually are, in order to analyze and find areas of energy waste and then to do calculations based on energy savings proposals.

PEAK HEATING, COOLING AND CFM PER AREA

☒ EXISTING ☐ NEW

BUILDING SUBURBAN OFFICE BUILDING DATE 11-18-1985

TOTAL SQ FT 90,000 AVERAGE CFM PER TON 400

AREA	SQ FT OF AREA	COOLING LOAD: SQ FT PER TONS	COOLING LOAD: TONS	HEATING LOAD: BTU PER SQ FT	HEATING LOAD: MBH	CFM SUPPLY: CFM PER SQ FT	CFM SUPPLY: CFM	DIRECT EXT. CFM
INTERIOR OFFICES, EAST	31,275	330	95	25	782	1.20	37,909	
INTERIOR OFFICES, WEST	31,275	330	95	25	782	1.20	37,909	
PERIMETER OFFICES, EAST	3,489	250	14	60	209	1.60	5,582	
PERIMETER OFFICES, SOUTH	4,950	225	22	60	297	1.60	8,800	
PERIMETER OFFICES, EAST	3,489	250	14	60	209	1.60	5,582	
CONFERENCE ROOM, EAST	450	250	2	60	27	1.60	720	1,200
CONFERENCE ROOM, WEST	450	250	2	60	27	1.60	720	1,200
VESTIBULE	225	100	2	35	8	4.00	900	
COMPUTER ROOM	3,600	140	26	35	126	2.90	10,286	
MENS TOILETS	224	400	1	53	12	1.00	224	225
WOMENS TOILETS	224	400	1	53	12	1.00	224	225
WOMENS LOUNGE	224	400	1	53	12	1.00	224	225
HALLWAY	300	400	1	35	11	1.00	300	
BUILDING ENGINEER	225	400	1	35	8	1.00	224	
STORAGE ROOM	1,800	650	3	35	63	0.61	1,108	
MECHANICAL ROOM	4,800	400	12	35	168	1.00	4,800	
KITCHEN	1,000	143	7	60	60	2.80	2,797	5,400
CAFETERIA	2,000	150	13	35	70	2.70	5,333	
TOTAL	90,000	290	310	31	2,883	1.37	123,642	8,475

Figure 2-16. Sample Filled Out Peak Heating, Cooling and CFM Per Area

ENERGY COST AND BTU PROFILES

Totals on energy costs and Btu energy consumption can be entered into a computer spreadsheet, and pies graphed out, to better see major or minor areas.

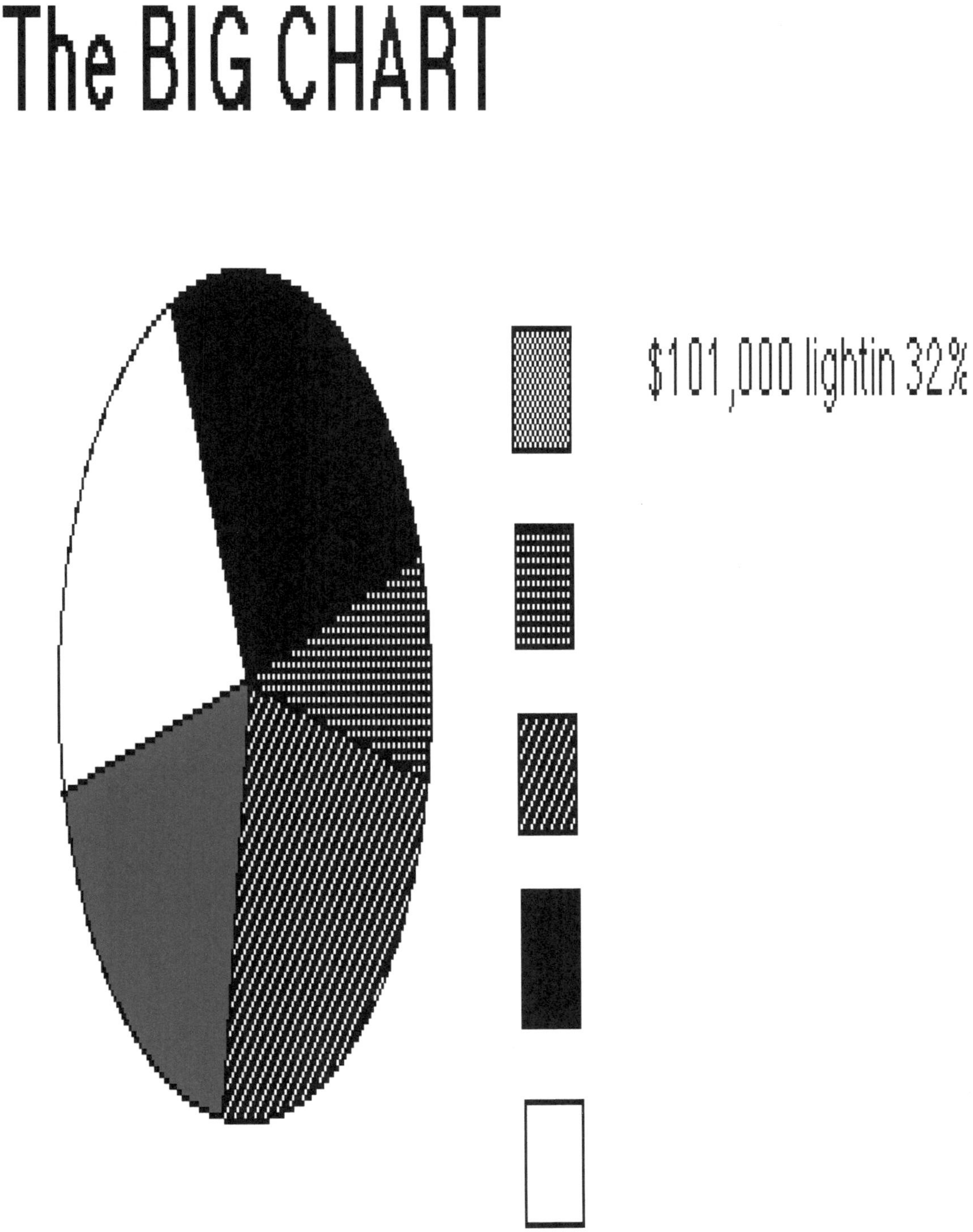

Figure 2-17. Sample Filled Out Energy Cost Profile/Btu Energy Consumption Profile

RESIDENTIAL AUDITS

The purpose of the residential energy audit forms is to help detect and evaluate unnecessary energy losses, and to help evaluate and select various energy savings.

The primary areas of energy problems in residences are as follows:

1. Lack of *insulation* in walls, ceilings, roof, floors, etc.
2. Lack of *storm windows* and doors, or double or triple pane windows.
3. *Low efficiency* heating and/or cooling equipment.
4. *Inefficient performance* from heating and/or cooling equipment. Inefficient combustion.
5. *Uninsulated ducts* or hot water pipes in unconditioned spaces such as the attic or crawl space.
6. Loose windows, doors, wall openings allowing excessive *leakage* in and out of the house.
7. No set-back/set-up *thermostats.* Improperly located stats.
8. *Unbalanced* systems with hot rooms and cold rooms.

ENERGY CONSUMPTION HISTORY

One of the most important things that has to be looked at in an energy audit is the energy consumption of the project.

In order to see if there is an energy waste problem or to what the extent there is of the problem—the heating fuel and electrical bills for the past year must be gathered and the amount of energy consumption and costs per each period listed.

Then the listing must be totaled for the year and the Btu per sq. ft. of building per year must be compared to other similar structures in the same area.

1. Fill out the heading first with the job name and type of heating fuel.
2. Fill in the period of time for each bill, number of days, unit amount of fuel used, convert to Btu and enter. Last, fill in the cost for the period.
3. Everything is converted to Btu's as the benchmark comparison of the totals.

Job ______________________ Date ______________________

ENERGY CONSUMPTION HISTORY

PRIMARY HEAT

☐ Natural Gas
Btu per cu. ft. ________

☐ Oil
Btu per gal. ________

☐ Propane
Btu per gal. ________

☐ Electricity
Btu per KW, 3412

☐ Coal
Btu per lb. ________

Fill in for a complete one year period

FUEL

Period	Days	Cubic Feet Gallons or LBS Consumed	BTU for Period*	Cost
Total For Year				

ELECTRIC

Period	Days	KW's Consumed	BTU for Period*	Cost
Total For Year				

*Multiply energy consumed by BTU per unit.
Example: 1000 cu. ft. gas x 1030 BTU/CU FT = 103,000 BTU

Wendes Engineering and Contracting Services Form AUD 302

Figure 2-18. Sample Filled Out Energy Consumption History Form

WALL, ROOF, FLOOR, WINDOW ENERGY EVALUATIONS

When conducting a residential energy audit it is often necessary to analyze the wall, roof, floor, window and door construction in detail and calculate the R factors and resultant U's.

These forms lead the auditor through the whole energy conservation calculation process from beginning to end and determine the bottom line value of the energy conservation proposals. They help determine:

Overall R and Resultant U Factors
Existing Conduction Heat Loss Per Year
Proposed Energy Improvements
New Heat Losses
Heat Loss Reduction Per Year
Fuel Reduction Per Year
Cost Savings Per Year

To determine the existing R factors and the factors for the energy conservation change, the components that make up the wall, etc. are listed, along with their individual conductivities, and a total R factor is calculated using ASHRAE or some other publication as a reference.

Job ____________________ Date ____________

OUTSIDE WALL ENERGY EVALUATION

List all materials including air space
Enter R factor and total.

	Thick-ness Inches	R Exist.	R New 1	R New 2
Interior Surface		.68	.68	.68
Surface				
Total R				
$\frac{1}{R_{total}}$ =	U			

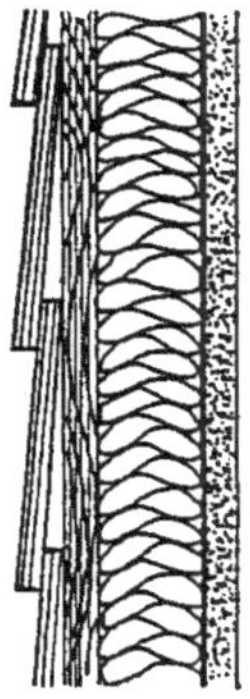

SQUARE FOOT AREA

EXISTING CONDUCTION HEAT LOSS PER YEAR

[______ x ______ x ______ x ______] ÷ ______ = [ ]
(U_x) (Net Area Sq Ft) (Avg Winter Temp Diff) (Total Winter Hours) (Divide by: Efficiency of Heating Equipment) (Heat Loss Mill. BTU Per Year)

PROPOSED ENERGY IMPROVEMENTS

Description ____________________

New Heat Loss: (Same Formula as Above) *New₁* *New₂*

[______ X ______ X ______ X ______] ÷ ______ = [ ]

[______ X ______ X ______ X ______] ÷ ______ = [ ]

Heat Loss Reduction Per Year — Subtract new heat loss from existing = [ ] [ ]

Fuel Reduction Per Year — Divide by BTU's per unit of fuel ☐Cu Ft ☐Therm ☐Gallon ☐KW → ÷ ______ ÷ ______ [ ] [ ]

Cost Savings Per Year — Multiply by ☐cost per unit of fuel or ☐cost per million BTU → X $ ______ per X $ ______ per [$] [$]

Wendes Engineering and Contracting Services — Form AUD 315

Figure 2-19. Sample Filled Out Outside Wall Energy Evaluation

Job ______________________________ Date ____________________

CEILING AND ROOF ENERGY EVALUATION

List all materials including air space
Enter R factor and total.

	Thick-ness Inches	R Exist.	New 1	New 2
Interior Surface		.68	.68	.68
Surface				
Total R				
$\frac{1}{R_{total}}$ =	U			

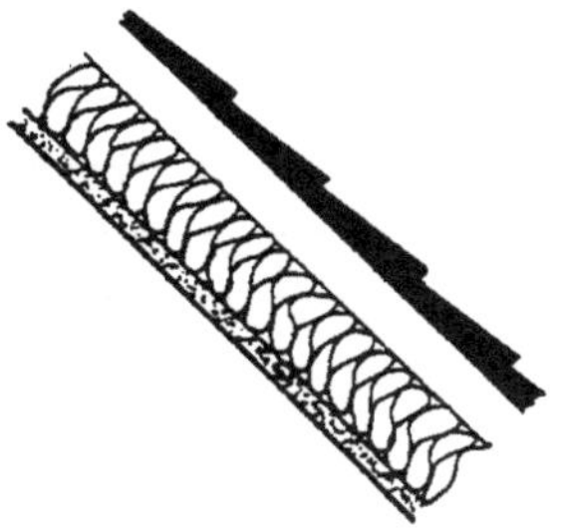

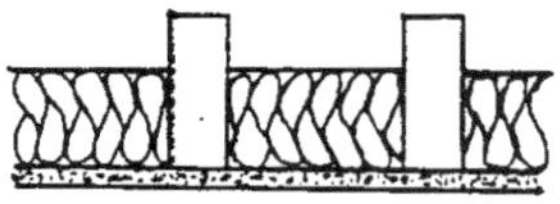

SQUARE FOOT AREA

EXISTING CONDUCTION HEAT LOSS PER YEAR

[______ x ______ x ______ x ______] ÷ ______ = [ ]

(U x) (Net Area Sq Ft) (Avg Winter Temp Diff) (Total Winter Hours) (Divide by: Efficiency of Heating Equipment) (Heat Loss Mill. BTU Per Year)

PROPOSED ENERGY IMPROVEMENTS

Description ____

New Heat Loss: (Same Formula as Above) New 1 New 2

[______ X ______ X ______ X ______] ÷ ______ = [ ]

[______ X ______ X ______ X ______] ÷ ______ = [ ]

Heat Loss Reduction Per Year — Subtract new heat loss from existing = [ ] [ ]

Fuel Reduction Per Year — Divide by BTU's per unit of fuel ☐ Cu Ft ☐ Therm ☐ Gallon ☐ KW → ÷ ______ [ ] ÷ ______ [ ]

Cost Savings Per Year — Multiply by ☐ cost per unit of fuel or ☐ cost per million BTU → X $ ______ per [$] X $ ______ per [$]

Wendes Engineering and Contracting Services

Form AUD 314

Figure 2-20. Sample Filled Out Ceiling and Roof Energy Evaluation

Job ________________________________ Date ____________________

CEILING AND ROOF ENERGY EVALUATION

List all materials including air space
Enter R factor and total.

	Thick-ness Inches	R Exist.	R New 1	R New 2
Interior Surface		.68	.68	.68
Surface				
Total R				
1 / Rtotal =	U			

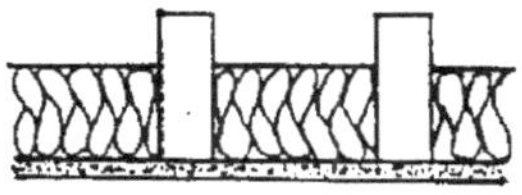

SQUARE FOOT AREA

EXISTING CONDUCTION HEAT LOSS PER YEAR

[______ x ______ x ______ x ______] ÷ ______ = [ ]

(U_x) (Net Area Sq Ft) (Avg Winter Temp Diff) (Total Winter Hours) Divide by: (Efficiency of Heating Equipment) (Heat Loss Mill. BTU Per Year)

PROPOSED ENERGY IMPROVEMENTS

Description ____

New Heat Loss: (Same Formula as Above) New 1 New 2

[______ X ______ X ______ X ______] ÷ ______ = [ ]

[______ X ______ X ______ X ______] ÷ ______ = [ ]

Heat Loss Reduction Per Year — Subtract new heat loss from existing = [ ] [ ]

Fuel Reduction Per Year — Divide by BTU's per unit of fuel ☐Cu Ft ☐Therm ☐Gallon ☐KW → ÷ ______ [ ] ÷ ______ [ ]

Cost Savings Per Year — Multiply by ☐cost per unit of fuel or ☐cost per million BTU → X $ ______ per [$] X $ ______ per [$]

Wendes Engineering and Contracting Services Form AUD 314

Figure 2-21. Sample Filled Out Windows and Doors Evaluation Form

Job ______________________ Date ______________

COLD FLOOR ENERGY EVALUATION

List all materials including air space
Enter R factor and total.

	Thick-ness Inches	R Exist.	R New 1	R New 2
Interior Surface		.68	.68	.68
Surface				
Total R				
$\frac{1}{R_{total}}$ =	U			

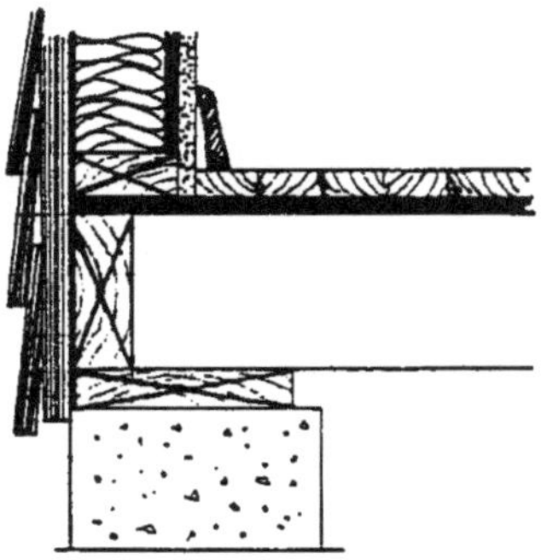

SQUARE FOOT AREA

EXISTING CONDUCTION HEAT LOSS PER YEAR

[______ x ______ x ______ x ______] ÷ ______ = [ ]

(U_x) (Net Area Sq Ft) (Avg Winter Temp Diff) (Total Winter Hours) Divide by: (Efficiency of Heating Equipment) (Heat Loss Mill. BTU Per Year)

PROPOSED ENERGY IMPROVEMENTS

Description ______________________

New Heat Loss: (Same Formula as Above) *New 1* *New 2*

[______ X ______ X ______ X ______] ÷ ______ = [ ]

[______ X ______ X ______ X ______] ÷ ______ = [ ]

Heat Loss Reduction Per Year — Subtract new heat loss from existing = [ ] [ ]

Fuel Reduction Per Year — Divide by BTU's per unit of fuel ☐Cu Ft ☐Therm ☐Gallon ☐KW → ÷ ______ ÷ ______ [ ] [ ]

Cost Savings Per Year — Multiply by ☐cost per unit of Fuel or ☐cost per million BTU → X $ ______ per X $ ______ per [$] [$]

Wendes Engineering and Contracting Services

Form AUD 317

Figure 2-22. Sample Filled Out Cold Floor Energy Evaluation

Chapter 3

Indoor Air Quality Procedures and Forms

PROBLEMS

Exposure to indoor air pollution is one of the major environmental health hazards in the United States today. IAQ problems lead to worker complaints, health problems, respiratory problems, reduced productivity and lawsuits. IAQ-related health problems are estimated at $15 billion a year and almost 2 billion lost work hours. About 30 percent of the people are complaining. Allergies caused by airborne biological particulates are estimated to effect 25 million people in the United States.

Indoor pollutant levels are often higher than those outdoors. Generally outdoor air is 20 times cleaner than indoor air and the average American spends 90 percent of his or her time indoors. Organic pollutants may be two to five times higher inside homes, and individual chemical pollutants up to 20 percent higher than outside.

Up to 20 percent of new commercial buildings may be affected by unhealthy concentrations of organic compounds as much as 100 times higher than found outdoors.

There is an imminent need for cleaner indoor air, to know and recognize causes of poor IAQ, to eliminate IAQ problems and pollutants, and to be able to diagnose, test and rectify effectively. Knowledge and skills with instruments, IAQ auditing procedures, analyzing skills, methods of correcting HVAC system deficiencies, etc., are required by those involved with IAQ.

SPECIFIC CAUSES OF INDOOR POLLUTION

— INADEQUATE VENTILATION
- Inadequate *Outside* Air
- Improperly *Controlled* Outside Air
- Inadequate *Exhausts*
  Toilet, Kitchen, Return Air, Fume, etc.
- *Tight* Building
- *Short Circuiting* of Air in Spaces
- Improperly *Pressurized* Building
- Inadequate *Filtering* of Air

— INDOOR GENERATED CONTAMINANTS
- Tobacco Smoke
- Gas Leaks
- Freon Leaks
- Aerosols, Cleaners
  Hair sprays, Cleaning Sprays, Disinfectants
- Pesticides
- Fumes, Chemical
- Nitrogen Dioxide
- Products of Combustion, Carbon Monoxide
  Holes in Heat Exchanger and Flues
  Clogged Flue or Chimney
  Engine Exhausts, car, trucks
- Copy Machines
- Lasers
- Hibachis and Charcoal Broilers

— BUILDING MATERIALS, FABRICS, FURNISHINGS
- Carpets
- Sheets and Blankets
- Carpet Adhesives
- Furniture
- Fabrics in Furniture and Drapes, etc.
- Wood, Plywood
- Insulation
- Paneling, Particle Board
- Plastics, Laminate
- Asbestos

— INDOOR BIOLOGICAL CONTAMINATION
- Standing Water
- Cooling Coils, Drain Pans
- Mildew Spores
- Carbon Dioxide CO_2 from People
- Pollen
- Mold
- Fungi
- Dust Mites
- House Dust
- Animal Dander
- Bacteria and Viruses
- Humidifiers (not evaporative type)

— OUTDOOR CONTAMINANTS
- Due to Infiltration
- Radon
- Soil Gas
- Methane
- Pesticides
- Auto Pollution
- Exhaust Stacks
- Due to Outside Air Intake Drawing in Contaminated Industrial Process

— INDUSTRIAL INDOOR CONTAMINANTS
- Paint
- Chemicals
- Printers
- Particulates, small solid or liquid particles such as dusts, powders, liquid droplets and mists. Examples: fly ash and asbestos dust.
- Gas Pollutants, fluids without form that occupy space rather than uniformly such as carbon monoxide or chloroform.
- Fumes, irritating smoke, vapor or gas.
- Pollutants may be toxic, noxious, corrosive erosive, inflammable, explosive or radio active.

TESTING, SETTING AND CONTROLLING OUTSIDE AIR

Correctly and consistently controlled minimum outside air volume is an absolute requirement for good IAQ and can be achieved as follows:

Outside Air Needed for IAQ

Provide sufficient outside air to meet codes, healthy ventilation, building pressure and direct exhaust makeup air requirements. Additional outdoor ventilation air has been shown to be the single most effective method of correcting and preventing problems and minimizing complaints related to poor indoor air quality. Even if a specific contaminant is identified (such as formaldehyde) dilution may be the most practical way of reducing exposures.

Residences:	0.35 air changes per hour This is comparable to 15 cfm per person.
Classrooms:	15 cfm per person
Offices:	20 cfm per person
Public Restrooms:	50 cfm per person
Where smoking is permitted:	60 cfm per person

This equates to generally 15 to 20 percent of total cfm for a 10,000 cfm system in a 10,000-sq.-ft office building which may need a minimum of 1,500 to 2,500 cfm. This equates to

.15 or .20 cfm per sq. ft. of building

The amount of fresh air naturally infiltrating into a tightly insulated home without ventilation may be as low as a hundredth (.01) of an air change per hour.

Check Out Actual Outside Air Conditions

1. Measure *actual amounts* of outside air being taken in under different outside temperature conditions and building load conditions.

2. Check if *minimum* air volumes are correct and being held. During occupancy periods, outdoor air dampers should not close beyond the minimum position and fans and air, handling units should run continuously.

3. During occupancy periods, outdoor air dampers should not close beyond the minimum position and fans and air handling units should run continuously. Check if correct amounts of OA are being taken in at *maximum* OA settings.

4. Check if sufficient amounts of cool *outside air* are being taken in when required for cooling.

5. Check winter, spring, summer and fall conditions.

6. Adjust *damper linkages* as required and reset *controls* or change controls as required.

7. Check that all supply outlets in spaces are opened to their correct balance positions.

What to Watch For

1. Don't reduce outside air volume below direct exhaust air quantities, thereby putting the building under a negative pressure and forcing air infiltration.

2. Maintain some building pressurization. This means bring in 1 to 5 percent more outside air than is exhausted.

Common Outside Air Plenums

If working with a common outside air plenum, put separate OA dampers on intake of each HVAC unit. The outside air plenum itself can be maintained as is.

CARBON DIOXIDE LEVELS

Measure CO_2 levels. They are a good indicator of the amount of fresh air being brought indoors and can help determine ventilation problems and needs.

ASHRAE recommends *indoor levels* of CO_2 kept below 1000 ppm
CO_2 levels increase with *occupancy* because people exhale CO_2
Outdoor levels of CO_2 typically range from 320 to 350 ppm

Infrared technology now makes the diagnosis of ventilation adequacy easy. Low cost ventilation efficiency measurement systems can quickly generate a ventilation record tracking CO_2 levels over time as a ventilation index.

Today's technology provides a number of possible low cost product solutions to ventilation adequacy, including ventostats (CO_2-based ventilation controllers), desiccant wheels, energy recovery equipment, etc. Demand control ventilation is one of a number of *control strategies* that can be used.

INDOOR AIR QUALITY INSTRUMENTS AND TESTING

There are a multitude of excellent testing instruments on the market, many for very reasonable prices. *Industrial Hygienists, HVAC Balancing Technicians* and *HVAC Engineers* may already own or be familiar with many of them.

- CARBON MONOXIDE, CO ANALYZERS
- CARBON DIOXIDE, CO_2 ANALYZERS
- RADON DETECTORS, RN
- MICROBIOLOGICAL SAMPLERS
- TOBACCO SMOKE TESTING
- VOLATILE ORGANIC COMPOUND, VOC SAMPLERS
- SEMI-VOLATILE ORGANIC COMPOUNDS
- BIOAEROSOL TESTING
- AIR FLOW MEASURING INSTRUMENTS SUCH AS AIR FLOW HOODS, PITOT TUBES AND ANEMOMETERS, HOT WIRE ANEMOMETERS, ETC.
- MOLD, YEAST, FUNGUS TESTS
- IAQ TESTING KITS
- TESTS FOR LEAD AND LEAD DUSTS
- TESTS FOR NICOTINE
- OZONE TEST
- TRACER GAS INSTRUMENTATION
- NATURAL GAS TEST
- TEMPERATURE RECORDERS
- HYGROMETER FOR HUMIDITY MEASUREMENTS
- REFRIGERANT LEAK DETECTORS
- TESTING FOR BIOAEROSOLS
- INSTRUMENTS FOR FORMALDEHYDE TESTING

HUMIDITY

The ideal humidity guideline should specify a relative humidity range that minimizes deleterious effects on human health and comfort as well as reduces, as much as possible, the speed of chemical reactions of the growth of biological contaminants (which will impact human health and comfort).

Like most gaseous and particulate contaminants, relative humidity is primarily affected by indoor and outdoor sources and sinks. However, unlike other contaminants, relative humidity is also a function of air temperature.

In addition to the effect of temperature, selecting the most desirable range of humidity is complicated by the conflicting effects of an increase or decrease in humidity levels. For example, while increasing humidity may reduce the incidence of common respiratory infections and provide relief for asthmatics, an increase in humidity may also increase the prevalence of microorganisms that cause allergies. Criteria for indoor exposure must balance both effects.

The bacterial population increases below 30% and above 60% relative humidity. The viral population increases at relative humidity below 50% and above 70%.

Fungi do not cause a problem at low humidity. However, their growth becomes apparent at 60%, increases between 80% and 90%, and shows a dramatic rise above 90%.

Mites require humidity for survival. Growth in the mite population responds directly to humidity levels in excess of 50%. The incidence of allergic rhinitis because of exposure to allergens increases at relative humidities above 60% and the severity of asthmatic reactions increases at relative humidities below 40%.

Most *chemical interactions* increase as the relative humidity rises above 30%, although ozone production is inversely proportional to the relative humidity.

The evidence suggests that the optimal conditions to enhance human health by minimizing the growth of biological organisms and the speed of chemical interactions occur in the narrow range between 40% *and 60%* relative humidity at normal room temperature. That narrow range is represented by the optimum zone in the shaded region of the graph.

Although keeping indoor humidity levels within this region will minimize health problems, there is probably no level of humidity at which some biological or chemical factor that affects health negatively does not flourish.

PARTICULATE CONTAMINANTS

PARTICULATES

Indoor air particulates may come from outdoor sources or indoor sources. Particulates are usually categorized according to size:

1. *Respirable,* less than 5 to 10 micrometers diameter which can lodge in the lungs and cause health problems.

2. *Nonrespirable,* greater than 5 to 10 micrometers diameter.

Typically reported range for total particulates of all sizes is 300 to 1000 micrograms per cubic meter averaged over 24 hours, with maximum readings of 600 micrograms per cubic meter. The indoor/outdoor ratio typically varies from 0.3 to 0.4. The following standards have been established:

TOBACCO SMOKE

Hypersensitivity to tobacco smoke is fairly common, often resulting in irritation of the eyes and respiratory tract.

Tobacco smoke consists of solid particles, liquid droplets, and gases, and constitutes more than 2,000 specific materials. Like many pollutants, it can be absorbed by the body unexpectedly. A study in Britain revealed traces of tobacco substances in the urine of a test group of nonsmokers in 85 percent of the cases, despite the fact that half of the group's members were unaware that they had been exposed to a low-level dose of tobacco smoke.

FUNGAL SPORES

Fungal spores are a broad class of biological organisms that can function as potential allergenic agents. Those most commonly found indoors are associated with mildew and decay and can be found in air conditioning systems. Fungal spores can also originate outdoors, where their numbers are subject to seasonal variations. Spore concentrations outdoors rarely exceed 1 spore/cc, and normal dwellings are likely to have lower concentrations.

FIBERS

Fibers can include several types of mineral or organic fibrous material. The most important of these from a health standpoint is asbestos, and only this type of fiber is considered here. Asbestos can occur in many forms, including amosite, chrysolite, and crocidolite. It may be found indoors through its use as a construction material (insulation), although this use has been severely curtailed in recent years.

Asbestos fibers, when lodged in the lung, can cause asbestosis (a disease of the lungs) and mesothelioma (a cancer that attacks the lining of the chest cavity or abdomen).

OSHA is using 0.1 fibers longer than five micrometers per cubic centimeter as the level above which abatement action must be taken.

The following standards have been established:

OSHA	2.0 fibers/cc (8-hour TWA for industrial exposure)
ACGIH	.5 to 2.0 fibers/cc (8-hour TWA for industrial exposure)
ASHRAE	lowest feasible level

PATHOGENS AND ALLERGENS

Pathogens are particulates that cause disease. Allergens, similarly, induce allergies. Buildings harbor both in the form of bacteria, viruses, mold spores, pollens, insect parts, and people and pet materials. Often, these particles adhere to other particles, primarily dust, and collect unseen in carpets, fabrics, duct systems, humidifiers, fan coil units, cooling towers, etc.

Exposure to such contaminants can induce a variety of allergic reactions and illnesses: cold, flu, and several forms of pneumonia, including Legionnaires' disease. Tuberculosis, measles, smallpox, *staphylococcus* infections, and influenza are known to be transmitted by air.

BIOLOGIC AEROSOLS

Increasing attention to biologic aerosol components of indoor air has resulted from investigations that have shown that airborne concentrations of viable organisms frequently correlate with physiologic responses and complaints. Symptoms including pulmonary manifestations, muscle aches, chills, fever, headache and fatigue have been attributed to biologic agents.

Disease has been attributed to thermophilic actinomycetes, non-pathogenic amoebae, fungi, and *Flavobacterium spp.* or their endotoxins.

FORMULAS

TLV: Stands for threshold limit value. It's the time-weighted concentration that normally healthy adults can withstand for eight hours a day (40 hours a week) without adverse effects. It is usually stated in parts per million (ppm).

TLV-STEL: The maximum concentrations to which workers can be exposed for up to 15 minutes.

TLVC: Is the level that cannot be exceeded even instantaneously.

Effective temperatures are sometimes used to express the effect that dry-bulb temperature, wet-bulb temperature, and air velocity have on comfort. It is a single-number index.

wbgt: Is the wet-bulb globe temperature, which includes the radiant effect.

DILUTION AIR

$$\text{cfm} = \frac{403 \times \text{SG} \times 10^6}{\text{MW} \times \text{TLV}} \times \text{pints/min} \times \text{k}$$

Where:

SG = specific gravity

MW = molecular weight

K = a safety factor that varies from 3 to 10 depending on the toxicity of the material and the effectiveness of the ventilation. This is difficult to estimate. We will use two values, K1 (the value for toxicity) and K2 (the value for ventilation), which together add up to K.

FIRE AND EXPLOSION

$$\text{cfm} = \frac{403\ (\text{S.G.})\,(100)\,(\text{C})\,(\text{pints/min})}{(\text{MW})\,(\text{LEL})\,(\text{B})}$$

Where:

LEL = Lower explosive limit (given as a percentage) which is expressed in parts per hundred.

MW = molecular weight of liquid

SG = specific gravity of liquid

C = safety factor, which depends on the percentage of the LEL necessary for safe operation because concentrations are not uniform. For continuous ovens, C=4, and for batch ovens, C=12.

B = a constant that takes into consideration the fact that LEL decreases at elevated temperatures. B=1 at temperatures less than 250°F, and 0.7 at temperatures equal to or more than 350°F.

INDOOR AIR QUALITY REPORT

Job ______________________ Date ______________

Location ______________________ Longitude ______________

VENTILATION

Building Under Pressure: ☐ Positve ☐ Negative ☐ Neutral

Building Pressure Readings, Inches S.P.: Inside ________ Outside ________ No. of Occupants: ________

Filters: Type ____________ Condition ____________

	Rated CFM	Actual CFM
Total Building Outside Air		
Total Building Exhaust Air		
Total Building Return Air		
Total Building Supply Air		
Infiltration		
Exfiltration		

CFM Outside Air Per Person ______________

Air Changes Per Hour ______________

CARBON DIOXIDE CO_2 READINGS

____

____

CARBON MONOXIDE CO READINGS

____

____

RELATIVE HUMIDITY RH READINGS

____

____

INDOOR CONTAMINANTS

____

____

OUTDOOR CONTAMINANTS

____

____

BIOLOGICAL CONTAMINANTS

____

____

INDUSTRIAL INDOOR CONTAMINANTS

____

____

Figure 3-1. Above is an overall report for indoor air quality pressure, ventilation and other contaminant tests and recording.

BUILDING VENTILATION AND PRESSURE REPORT

Job ______________________________ Date ______________

Location ______________________________ Longitude ______________

Size of Building, Sq Ft ______________________________ No. of Floors ______ No. of Bldgs. ______

Building Under Pressure: ☐ Negative ☐ Positive ☐ Neutral

Pressure Readings: Inside _____ Outside _____ Floor _____ Walls: N______ S_______ E______ W______

Time: ☐ Winter ☐ Spring ☐ Summer ☐ Fall ☐ Daytime ☐ Night

Outside Temperature ________ Inside Temperature ________ ☐ Occupied ☐ Unoccupied

System Tag	Type System	Location In Building	☐ Outside Air ☐Recirc Air ☐ Exhaust Air[1] ☐Supply Fan[2]					
			Maximum CFM		Minimum CFM		Random CFM	
			Rated	Actual	Rated	Actual	Rated	Actual

Door, Window, Building Opening Conditions ______________________________

____

Stack Effect ______________________________

____

Wind Effect ______________________________

____

Combustion Effect ______________________________

____

Remarks ______________________________

____

____

____

1: Exhaust air to outside 2: Supply fan discharge CFM

Figure 3-1. Above is detailed report on ventilation per system.

Chapter 4

Engineering Calculation Procedures and Forms

This chapter has the engineering calculation forms for heat loss and cooling load calculation for either peak loads or seasonal totals.

It also has a summary sheet for listing the peak heating and cooling loads and cfm per area.

HEAT LOSS CALCULATION FORMS

The original heat loss calculations and capacity HVAC equipment chosen at the time the building was constructed may have greatly changed over the years and new heat loss calculations may be required, based on the actual current conditions' in order to end up with a reliable energy retrofit program.

Lighting loads may have been reduced, insulation added, thermal windows added, and temperature settings changed, etc. since the building was built. The old figures may no longer be valid.

New calculations based on the current conditions are required for a valid audit and energy conservation program.

1. The heading on the heat loss calculation sheet describes the building, its size, design temperatures, etc.
2. These lines list all the components involved in the heat loss calculations:
   Roofs
   Floors
   Walls
   Windows, etc.
3. The middle columns are for the dimensions, sq. footage, U factors and temperatures differences.
4. Multiply the *square feet* times *the U factors* times *the temperature difference* for the total Btuh for peak load per hour.
5. Use appropriate formulas for equipment, motors and ventilation.
6. The far right column lists the seasonal hours involved.
7. The total peak Btu's are summed at the bottom.

HEAT LOSS CALCULATION

☐ PEAK PER HR ☒ SEASONAL ☒ EXISTING ☐ NEW

Building SUBURBAN OFFICE BUILDING Date 11-18-1985

Location Latitude 41

Type Building OFFICE AND LABS Stories 1 When Built 1967

Sq Ft Area 90,000 Cubic Ft of Space 1,260,000

X Calculation for Whole Building For Partial Area

Budget Load: Sq Ft 90,000 x BTUYR/Sq Ft 78,500 = 7,065 Mill BTUYR

Outside Design: DB -10 WB Avg OA Temp. 35 Winter Hours 4,800

Inside Design: DB(day) 74 RH DB(night)

ITEM	DIMENSIONS	SQ FT	U	TEMP DIFF	BTU: Per Hour Seasonal	SEASONAL HOURS
ROOF OR CEILING	360 X 250	90,000	0.110	39.00	1,853,280,000	4,800
FLOOR	360 X 250	90,000	0.100	16.00	691,200,000	4,800
GLASS	950 X 7	6,650	1.100	39.00	1,369,368,000	4,800
DOORS						
WALLS	1220 X 14	10,430	0.330	39.00	644,323,680	4,800
COLD INSIDE WALLS						
VENTILATION	1.43 AC/HR CFM=	30,000	1.080	39.00	6,065,280,000	4,800
DUCT LOSSES						
GROSS TOTAL HEAT LOSS					10,623,451,680	
HEAT GAINS; LITES	4W/SQ FT	90,000	4.000	3.416	3,074,400,000	2,500
PEOPLE	No. People =	400	250		200,000,000	2,000
OFF. EQUIPMENT	HP =	25	2,550		127,500,000	2,000
COMPUTERS	KW =	5		3416.00	149,620,800	8,760
TOTAL INTERNAL GAINS					3,551,520,800	
NET TOTAL BUILDING HEAT LOSS					7,971,930,880	78,577*
INPUT TO HEATING EQUIPMENT, efficiency =			.60		11,800,000,000	131,110*

BTUH = Sq Ft x U x Temp. Diff. BTU/YEAR = Sq Ft x U x Avg Temp. Diff. x Winter Hours

* BTU/SQ FT/YEAR

Remarks____________________

Wendes Mechanical Consulting Services Co. Form AUD 307

Figure 4-1. Sample Filled Out Heat Loss Calculation Form

COOLING LOAD CALCULATION FORM

The original cooling load calculations and capacity HVAC equipment chosen at the time may have greatly changed over the years, and new cooling load calculations may be required based on the actual current conditions, in order to end up with a reliable energy retrofit program.

Lighting loads may have been reduced, insulation added, thermal windows added, temperature settings changed, etc., since the building was built. The old figures may no longer be valid.

New calculations based on the current conditions are required.

1. The heading on the cooling load calculation sheet describes the building, its size, design temperatures, etc.

2. The lines below list all the components involved in the cooling loads calculations:
   Roofs
   Floors
   Walls
   Windows, etc.

3. The middle columns are for the dimensions, sq. footage, U factors and temperatures differences.

4. Multiply the *square feet* times *the U factors* times *the temperature differences* for the total Btuh for peak load per hour.

5. Total peak Btu's and tonnage are summed at the bottom.

COOLING LOAD CALCULATION

☒ PEAK PER HR ☐ SEASONAL ☒ EXISTING ☐ NEW

Building: SUBURBAN OFFICE BUILDING Date: 11-18-1985

Location: Latitude: 41 Peak Load, HR, MO:

Type Building: OFFICE AND LABS Stories: 1 When Built: 1967

Sq Ft Area: 90,000 Cubic Ft of Space: 1,260,000

X Calculation for Whole Building For Partial Area

Building Load: Sq Ft 90,000 x Sq Ft/Ton 270 = 333 Tons

Outside Design: DB 95 WB 75 Avg OA Temp. 35 Summer Hours 900

Inside Design: DB(day) 74 RH DB(night) 74

ITEM (Orient.)	DIMENSIONS	SQ FT	U or FACTOR*	TEMP DIFF	SENSIBLE BTUH	TONS	LATENT BTUH
ROOF OR CEILING		90,000	0.110	20.00	198,000	16.50	
GLASS EAST		1,500		15.00	22,500	1.88	
(Solar) SOUTH		2,160		50.00	108,000	9.00	
WEST		1,500		15.00	22,500	1.88	
GLASS SOUTH		5,160	1.130	20.00	116,616	9.72	
(Conduction)						0.00	
WALLS E,W,N		9,040	0.167	20.00	30,194	2.52	
(Conduction)		2,880	0.167	50.00	24,048	2.00	
LIGHTING		90,000	4	3.416	1,229,760	102.48	
PEOPLE, NO.	(NO.)	400	250	250	100,000	8.33	100,000
OFF. EQUIPMENT	MOTORS (HP)	25	2,550		63,750	5.31	
COMPUTERS	(KW)	10	3,416		34,160	2.85	
KITCHEN						0.00	
VENTIL. AIR, Sens	1.43 AC/HR	30,000 CFM	1.08	20.00	648,000	54.00	
VENTIL. AIR, Lat.	1.43 AC/HR	30,000 CFM	0.010	4,840		0.00	1,452,000
TOTAL COOLING LOAD,			SENSIBLE		2,597,528	216	
			LATENT		1,552,000	121	
			GRAND TOTAL		4,149,528	337	267*

BTUH = Sq Ft x U x Temp. Diff.
* U, CLTD, CLF, KW, HP, WD
CLTD = (Temp diff) - (Daily Range + 14)/2
BTU/YEAR = Sq Ft x U x Avg Temp. Diff. x Hrs

* Sq Ft/Ton

Wendes Mechanical Consulting Services Co. Form ENG 402

Figure 4-2. Sample Filled Out Cooling Load Calculation Form

PEAK HEATING, COOLING AND CFM PER AREA

This very valuable form shows the peak heating and cooling loads for the main areas in the building and the amount of cfm in each area.

The form can be used to examine the original design or what the current loads actually are, in order to analyze and find areas of energy waste. Then do calculations based on energy savings proposals.

PEAK HEATING, COOLING AND CFM PER AREA

☒ EXISTING ☐ NEW

BUILDING SUBURBAN OFFICE BUILDING DATE 11-18-1985

TOTAL SQ FT 90,000 AVERAGE CFM PER TON 400

AREA	SQ FT OF AREA	COOLING LOAD		HEATING LOAD		CFM SUPPLY		DIRECT EXT. CFM
		SQ FT PER TONS	TONS	BTU PER SQ FT	MBH	CFM PER SQ FT	CFM	
INTERIOR OFFICES, EAST	31,275	330	95	25	782	1.20	37,909	
INTERIOR OFFICES, WEST	31,275	330	95	25	782	1.20	37,909	
PERIMETER OFFICES, EAST	3,489	250	14	60	209	1.60	5,582	
PERIMETER OFFICES, SOUTH	4,950	225	22	60	297	1.60	8,800	
PERIMETER OFFICES, EAST	3,489	250	14	60	209	1.60	5,582	
CONFERENCE ROOM, EAST	450	250	2	60	27	1.60	720	1,200
CONFERENCE ROOM, WEST	450	250	2	60	27	1.60	720	1,200
VESTIBULE	225	100	2	35	8	4.00	900	
COMPUTER ROOM	3,600	140	26	35	126	2.90	10,286	
MENS TOILETS	224	400	1	53	12	1.00	224	225
WOMENS TOILETS	224	400	1	53	12	1.00	224	225
WOMENS LOUNGE	224	400	1	53	12	1.00	224	225
HALLWAY	300	400	1	35	11	1.00	300	
BUILDING ENGINEER	225	400	1	35	8	1.00	224	
STORAGE ROOM	1,800	650	3	35	63	0.61	1,108	
MECHANICAL ROOM	4,800	400	12	35	168	1.00	4,800	
KITCHEN	1,000	143	7	60	60	2.80	2,797	5,400
CAFETERIA	2,000	150	13	35	70	2.70	5,333	
TOTAL	90,000	290	310	31	2,883	1.37	123,642	8,475

Wendes Mechanical Consulting Services Co. Form ENG 403

Figure 4-3. Sample Filled Out Peak Heating, Cooling and CFM Per Area

Chapter 5

HVAC Cost Estimating Procedures and Forms

The forms in this chapter will make estimating for HVAC work more accurate, clearer and complete. They will help save time and money and avoid errors.

Forms are an indispensable aid and guide to efficient estimating:
— They help control the proper sequence of estimating work.
— They continually remind you of what information is needed.
— They lead you logically through calculations.

As a result your bids will be more complete and correct.

For those who normally are not directly involved with HVAC work, the forms will help to clarify what's involved in HVAC estimating.

The forms are excellent for checking change orders too.

REQUIREMENTS OF A GOOD SHEET METAL ESTIMATOR

Solid estimates are produced by competent and reliable estimators. Good sheet metal estimators are developed through the following background of knowledge, procedures, skills and abilities:

Estimating Principles and Procedures

1. Estimators must follow sound efficient *procedures* for preparing estimates, such as:
   a. Become thoroughly *familiar* with the project, the types of systems and ductwork involved in the scope of work, before starting a detailed takeoff.
   b. Be familiar with *budget* estimating: HVAC costs for different buildings based on cost per sq. ft. of building or costs per ton of air conditioning; amount of ductwork per sq. ft. of building or by the average size; cost of ductwork per linear ft, per lb or per sq. ft.
   c. Know the *major divisions* of the estimate:
      Equipment
      Ductwork
      Duct Accessories, Sheet Metal Specialties
      Special Labor
      Sub-Contractors
      End of Bid Factors (such as sales tax)
      Markups for Overhead and Profit
   d. Be familiar with *detailed scope* of what is required in a sheet metal estimate.
   e. *Highlight* drawings before the takeoff.
   f. Follow systematic overall *procedure*
      Study the plans and specs
      Send out quotation requests
      Highlight drawings
      Make takeoffs and extensions
      Summarize
      Recap and markups
   g. Do constant systematic *checking* on each part as you go along and overall at the end. Double check everything.

2. Estimators must have the ability to *read blue prints,* recognize symbols, types of ductwork, equipment and systems, etc.

Air Distribution Systems

3. Estimators must be familiar with the different *types of HVAC* systems such as:
    - Low Pressure Constant Volume Systems
        - Single Zone, Reheat Coils, Multi-Zone
    - High Pressure Constant Volume
        - Dual Duct, Induction, Reheat Terminal
    - Variable Air Volume
        - Cooling Only, Cooling/Reheat Terminals,
        - Fan Powered, Dual Duct
        - Induction, Multi-Zone
        - System Powered, Riding Fan Curve
        - Damper Terminal By-pass
    - Exhaust Systems
        - Return Air, Toilet
        - Kitchen, Lab, Industrial

    They must not only recognize the various types of systems on plans, but must know all of the *components* required in them, whether shown on the plans or not.

4. They must know *duct pressure* and velocity ranges:

    Duct Pressure Ranges: 1/2 inch, 1 inch, 2 inch
    3,4,5 inches
    6 inches and up

    Velocities 0 to 2,000 fpm.
    2,000 fpm. and up etc.

5. They must know about different ductwork *system configurations* such as:
    - Single Duct
    - Dual Duct
    - Multi-Zone
    - Loops
    - Plenum Ceilings

6. They must have some familiarity with air distribution system *design,* know recommended air speeds, pressure drops and duct sizes plus sizing and selection of equipment.

Types of Ductwork

7. Sheet metal estimators must be familiar with the different *types of ductwork* and of their correct construction.
   - Rectangular Galvanized: Low, Medium, High Pressure
   - Low Pressure Round Ductwork; Flues; Flexible Tubing
   - Spiral Pipe and Fittings
   - Fiberglass Ductboard
   - Light Gauge Aluminum, Stainless, PVS with cleats, Pittsburghs
   - Heavy Gauge Metals: Black Iron, Stainless, Aluminum, Galvanized, Corten, etc.
   - PVC, FRP, Sunstrand

8. They must know the correct *applications* of different types of ductwork materials to various systems:
   - Low, Medium, High Pressure HVAC Systems
   - General Exhausts
   - Fume Exhausts
   - Heat Exhaust Systems
   - Chemical Exhaust Systems
   - Abrasive Material Systems

Ductwork Construction

9. Estimators must be familiar with the different types of *connections* for each type of ductwork and of their correct application to different types of systems.
   - Cleats: drive, flat S, bar, reinforced bar
   - Transverse: TDC, TDF
   - 4 Bolt Connections
   - Angle Flange, Van Stone
   - Bent Angle Flange
   - Butt Welded
   - Slip, Couplings

10. They must be familiar with different types of *seams* used for constructing ductwork.
    - Pittsburgh
    - Snaplock, Lockseam
    - Welded

11. They must be familiar with the different *gauges* **used for ductwork and specialties.**
    - Commercial Galvanized 26 to 16 gauge
    - Residential Galvanized 30 to 18 gauge
    - Heavy Gauge Industrial 18 gauge to 1/2 inch thick
    - Fiberglass Ductboard, 1 inch thick
    - PVC, 1/4," 3/16" inch thick

12. They must be familiar with the different types of *reinforcing* used on ductwork.
    - Angles
    - Channels
    - Crossbreaking
    - Tie Rods

13. They must be familiar with the different types of *fittings* used in air distribution systems and their correct applications.

Elbows:	Radius Throat, Square Throat
	90, 45. 22 1/2 degree etc.
Transitions:	Equal Taper, FOT, FOB, sq. side etc.
Offsets:	Ogee, Square
Wye Fittings	
Tap In Tees	

Estimating Materials and Labor

14. They must know the various *methods* of estimating ductwork.

15. Sheet metal estimators must know how to estimate ductwork *materials.*
    - Takeoff and calculate surface sq. footage of material based on size, length, etc.
    - Add waste and seam factor
    - Multiply by weight per sq. foot

16. They must be familiar with different waste and *allowance factors* for seams, cleats, hangers, hardware, etc.

17. They must know the methods of estimating ductwork *labor* such as:

Per Piece	Per Batch
Per Pound	Per Break Down of Component Parts
Per Sq. Foot	
Per Linear Foot	

They must know *sources of labor* such as the Wendes Sheet Metal Estimating Manual, cost records, etc.

Correction Factors

18. They must apply *labor multipliers* with reasonable accuracy whenever needed to adjust for conditions etc.
    5th floor takes 10 percent longer
    30 ft. high ductwork takes 20 percent longer
    Duplicate fittings go 33 percent faster

Accessories

19. He must be familiar with the various duct *accessories* and *sheet metal specialties.*
    Turning Vanes Air Foil, Single Skin
    Splitter Dampers
    Single and Multiblade Dampers
    Access Doors

Fabrication and Installation Procedures

20. They must be familiar with *fabrication procedures* and machinery and how those affect labor and overhead margins. They must be familiar with plasma cutters, coil lines, seam machines, press breaks, rollers, etc.
    Plasma cutters cut overall fitting labor in half
    Duct coil lines reduce straight duct labor about 70 percent

21. Sheet metal estimators must be well versed in ductwork *installation procedures,* in the operations involved in installations, with hand tools, scaffolding, Vermets, and scissor hoists.

Pricing Equipment

22. They must know *sources of pricing* on accessories and equipment, supplier, price catalogues, suppliers for quotations, etc.

23. They have to know about *small ventilation equipment:*
    Grilles, Registers, Diffusers
    Multiblade Dampers, Back Draft Dampers
    Fire Dampers, Access Doors

24. They must know about sheet metal *specialties* such as:
    Sheet Metal Housings, Walk Through Doors
    Belt Guards, Drain Pans, Coil Stands
    Coil Blankoffs

25. They must know about major HVAC *equipment.*
    Roof Top Units, Air Handling Units
    Fans, Filters, Louvers

Wage Rates, Unions, Jurisdictions

26. They must know about *wage rates,* union fringe benefits, federal, state and local taxes, insurance, etc.

27. They must be knowledgeable about union, trade and local *labor jurisdictions.*

28. They must be familiar with building *codes.*

Other Trades, Types of Buildings

29. They has to be familiar with *other trades* such as piping, insulation, temperature control, electrical and excavation.

30. They must be familiar with all types of *buildings,* commercial, institutional, industrial, their general sizes, layout, etc., and with the sequence of general construction work.

Markups

31. Sheet metal estimators must be generally familiar with financial statements such as profit, loss and balance sheets. They must be able to determine the correct markup for overhead and profit for their company and for the particular job they are bidding.

    They should understand how overhead costs are *pro-rated* onto direct material and labor costs for different projects, for different levels of sales and overhead costs, for different ratios of material to labor etc.

Skills, Traits Required

32. Estimating requires a host of skills, mathematical, mechanical, reading, writing, visualizing and drawing. It requires being methodical, analytical, strategical and realistic.

33. It absolutely demands that estimators be *reliable,* that They be thorough in their understanding of the project, of its scope, in takeoffs, interpretations, extensions, summaries, recaps.

Thus, knowledgeable, proficient and reliable estimators as described above will be able to produce complete and accurate estimates, which in turn become the required foundation blocks of successful contracting.

EFFICIENT ESTIMATING PROCEDURES

The following diagrammatic procedure is an efficient, systematic, organized, time saving approach for controlling the preparation of your bids.

- It promotes efficiency You get your bids **done faster**. You avoid duplications, and redoing of work, and you get things done simultaneously.
- You are more likely to get your bids **done on time** and thereby allow time for proper checking and solving of problems. You avoid hectic 11th-hour scrambling.
- **It provides framework** for planning and scheduling estimating work realistically and effectively.
- Promotes **more complete and accurate bids**, thorough takeoffs, accurate extensions and reliable pricing.
- Through the systematic ten-phase pyramidal procedure you will **produce more estimates,** reduce errors and get more jobs.

PRELIMINARIES

This first phase of the procedure is the most important step of all and it sets the ground work for a proper bid.

The preliminary surveys are a systematic, scientific, highly organized approach to becoming **thoroughly familiar** with a job before preparing an estimate and getting into the quagmire of details.

In the preliminary survey you study the plans, specs and other documents, to become familiar with what is involved in the project, what the scope is, what is included and not, what the approximate budget price is, the size of the building and what rough quantities of metal and equipment there are.

You determine if there are alternates or addenda and what the bidding instructions are.

You become familiar with the areas, floors, systems, equipment, ductwork, conditions, specialties, subs, etc.

You evaluate the competition, AE, generals, agencies and inspectors involved, cash flow, your work load, the construction schedule, your ability and experience to do the job, your competitive stance, and amount of time to bid the job. They you determine intelligently and realistically if you should bid the job or not.

Finally, you use the preliminary survey as your note sheet and checkoff list.

Part of the preliminary works is to notify subcontractors and equipment suppliers from whom you will be needing a quotation for the particular job. Give them the date it is needed so they have adequate time to prepare and can work simultaneously.

Also, make arrangements for any forms needed, pre-qualifications, written proposals, bid bonds, bid deposit checks, etc., so that they are ready at the bid time.

Quantity Takeoffs and Extensions

Before beginning the takeoff of ductwork and equipment, thoroughly study the plans and specs and mark and color the drawings. Highlight different types of ductruns, lined ducts and insulated ducts in color as required to distinguish one from the other.

Locate and mark alternate and addendum areas and conditions that require labor adjustments. Take off ductwork first, then specialties and finally equipment.

Summary Sheet

Extend materials, costs, labor, etc. on the takeoff sheet up and transfer totals of categories to summary sheet. List everything on the summary sheet, grouping items in the major categories; equipment to start with, then ductwork, specialties, special labor and minor subs.

Price out raw materials, extend shop and field labor and total the labor columns.

Quotations

Call for the quotations that have not come in yet. Make sure they have essential information on them such as quantities, types, manufacturers, accessories, exclusions, delivery; do they meet plans and specs, and are materials, sizes, performance correct?

Organize and compare the quotations and select the lowest acceptable ones. Plug numbers into summary sheet and total material column.

Make Thorough Check

Make a thorough check now of everything you have done to this point. Check all takeoffs, extensions, summations, transferences, pricing, labor, etc. Have someone else study project and review your estimate. Reread plan, specs, notes, quotes, etc. Have someone else check the math.

Recap

Transfer correct totals from summary sheet to the recap sheet. Price out labor and summarize subs. Put in end-of-bid factors such as sales tax, bonds, material and labor increases. Determine the proper markup for overhead and enter. Add everything together and add the desired profit to it. Check over recap carefully because errors can be costly.

Bid Submission

Submit a proper, qualified bid noting inclusions, exclusions, and exceptions to plans and specs.

ESTIMATING PROCEDURE DIAGRAM FOR PIPING AND SHEET METAL WORK

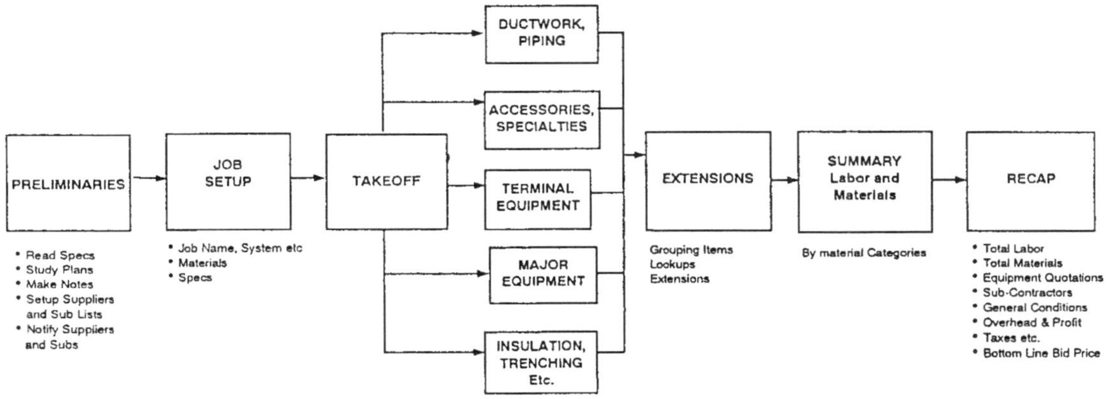

Figure 5-1. The above diagram shows a complete, fast and efficient procedure for preparing sheet metal estimates. The diagram shows the correct sequence of operations and the main areas of work. It follows the critical path method showing the sheet metal estimator, HVAC equipment supplier and sub-contractor all preparing their own portions of the estimates at the same time and all coming together for a total bid price within the bid time frame.

SCOPE OF COMPLETE SHEET METAL ESTIMATE

AIR HANDLING EQUIPMT

HVAC Units
- Air Handling Units
- Coils, Filters, Dampers
- Roof Tops, Economizers
- Make Up Air
- Heat Recovery

Fans
- Centrifugal
- Vent Sets
- Tubular
- Industrial
- Propeller
- Isolators
- Inlet Vanes
- Drives, Motors
- Access Doors

Roof Equipment
- Roof Exhaust Fans
- Gravity Vents
- Louvered Vent Houses
- Roof Curbs

Filters
- FG Throw Away
- Washable Mesh
- Rollomatic
- Electro Static
- Bag
- Absolute
- Charcoal

Air Diffusion Eqpmt
- Grilles
- Diffusers, Slot Diffusers
- Linears,
- Troffers
- Extractors

Terminal Eqpmt
- VAV Boxes
- Dual Duct
- Reheat
- Induction

Industrial Exhaust Eqpmt
- Auto, Welding Exhausts
- Dust Collectors
- Scrubbers
- Paint Spray Booths

Dampers
- Manual
- Motorized
- Fire, Smoke

Electrical
- Starters
- Heating Coils

Factory Fab. Housings

Sound Attenuators

Thermometers

DUCTWORK

Galvanized
- Low, Medium, High Pressure

Fiberglass

Round Oval HVAC
- Spiral
- Flex Tubing
- Furnace Pipe
- Flues

Industrial Exhaust
- Black Iron
- Stainless Steel
- Aluminum
- Blow Pipe
- FRP
- PVC
- PVC Coated Galvanized
- Metal Flex. Tubing

SHEET METAL SPECIALTIES

Duct Accessories
- Turning Vanes
- Splitter Dampers
- Bracing Angles
- Cleat, Hangers
- Trapeze Angles
- Spanning Angles
- Crossbreaking
- Seal Ducts
- Paint Ducts

Specialties

Flexible Connections
Belt Guards
Hsg. Access Doors
Hoods-Kit Lab, Shop
Stands, Platforms
Drain Pans
Blankoffs, Safeoffs
Roof Hoods
Screens, Grating
Expanded Metal
Perforated Plates
Water Elimination
Lead, Cork, Foam Glass

Lining
- F.G. Flexible
- Hardboard
- Cement and Pins

Sheet Metal Housings

Panels
Angles

SPECIAL LABOR
- Shipping
- Field Measuring
- Drafting
- Testing and Balancing
- Leak Testing
- Temporary Heat
- Set Up and Clean Up
- Chases and Sleeves
- Existing Buildings
  - Removal
  - Cut Openings, Patch
  - Protection

SUB CONTRS, RENTALS
- Testing and Balancing
- Insulation
- Temperature Control
- Cranes, Hoists
- Concrete Pads
- Scaffolding
- Refrigeration, AC
- Heating
- Electrical
- Cut Openings, Patch
- Excavate, Backfill

END OF BID FACTORS
- Sales Tax
- Permits
- Travel Pay
- Room and Board
- Wage Increases
- Material Increases
- Premium Time
- Alternates
- Contingencies
- Clean Up Charges

MARKUPS

Overhead
Profit

HVAC ESTIMATE

Date ______________________

Job ____

Bid ______________________ Time ______________________

Place ____

Estimator ____

Architect ______________________ Phone ______________________

Engineer ______________________ Phone ______________________

Figure 5-2. Coversheet For HVAC Estimate Report

JOB DESCRIPTION AND BUDGET COSTS

The purposes of this form are:

1. To budget estimate prices to determine if it should be bid or not and as a check price against the detailed estimate.

2. To approximate heating, cooling and cfm loads and roughout ductwork weight for check on detailed ductwork takeoff.

3. To record the key aspects of the types of systems involved.

JOB DESCRIPTION AND BUDGET COSTS

Job Zayres Date Feb. 9, 1987

Location 135 Busse Rd. Distance 10 Miles

Total Project Costs $ 2,240,000 Volume of Building — Cu Ft

Total Area 80,000 Sq Ft, $Area_1$ ______ Sq Ft $Area_2$ ______ Sq Ft

BUDGET COSTS

	COST/SQ FT BLDG	TOTAL	COST/TON	TOTAL
Total HVAC	$ 3.00	$240,000	$ 1,053	$240,000
Sheet Metal	$ .60	$ 48,000	$ 210	$
Piping	$ .07	$ 5,400	$ 23	$
Equipment	$ 2.14	$171,000	$ 754	$
Insulation	$ .17	$ 13,300	$ 58	$
Temperature Control	$ .03	$ 2,400	$ 10	$
Electric	$ —	$ —	$ —	$

DESIGN LOADS

	$Area_1$		$Area_2$	
	Factor	Total	Factor	Total
Cooling	350 Sq Ft/Ton	228 Tons	Sq Ft/Ton	Tons
Cooling	34 BTU/Sq Ft	2,720,000 BTU	BTU/Sq Ft	BTU
Heating	32 BTU/Sq Ft	2,560,000 BTU	BTU/Sq Ft	BTU
Supply Air	1.1 CFM/Sq Ft	88,000 CFM	CFM/Sq Ft	CFM
Duct Weight	.30 LBS/Sq Ft	24,000 LBS	LBS/Sq Ft	LBS

SYSTEM DESCRIPTION

Heating	Gas
Cooling	DX
☑ SZ ☐ MZ ☑ Constant Volume ☐ VAV	
Duct Pressure	☑ LP ☐ MP ☐ HP
Return Air Method	☐ Duct ☑ Ceil. Plenum
Type Outlets	Diffusers
Type Perimeter Heat	☑ Air ☐ Baseboard
Temperature Control	Electric

KEY PLAN

No. of Buildings ______ Stories ______

CONSTRUCTION: Class Thermopane Gross Area 3,300 Sq Ft; U .50

Exterior Walls 12" Masonry Gross Area 18,550 Sq Ft; U .10

Roof 2" insulation, metal deck Gross Area 80,000 Sq Ft; U .067

Remarks ______

Wendes Engineering and Contracting Services 8/80 Form EST 101

Figure 5-3. Sample Filled Out Job Description and Budget Costs Form

BID CHECKOFF SHEET

The purpose of the bid checkoff sheet is to make organized notes of everything required on a bid as a guide in preparing the estimate and then as a checkoff list to make sure everything is covered at the end of the bid.

The form is divided up into the major categories of an estimate, equipment, ductwork, piping, specialties and accessories, special items, subcontractors, rentals and end-of-bid items.

The form also covers other pertinent items in the bidding and pricing process and items not included in the bid.

BID CHECK OFF SHEET

☐ Total HVAC ☐ Sheet Metal ☐ Piping

Job Zayre's Discount Store Due Date Feb 28, 1987

Location ________ Distance ________

Project Costs $2,240,000 Size, Sq Ft 80,000 Stories 1

Budget Estimate ________ Buildings 1

EQUIPMENT	☐ DUCTWORK ☐ PIPING
6 Roof Top Units, 38 Tons ea.	Galvanized, Low Pressure, SMACNA
Gas, DX, Economizers	BI Kitchen Exhaust, 16 Ga.
2 Roof Exhaust Fans	
2 Prop Fans with Shutters	
20 Diffusers	
7 Registers	
20 Egg Crate Grilles	
3 Grease Filters	
	SPECIALTIES
	Turning Vanes
	Splitter Dampers
	INSULATION
	1" Thick, 1½ lb Lining
	Supply Ductwork

Wendes Engineering and Contracting Services 7/80 Form EST 102

Figure 5-4. Sample Filled Out Bid Checkoff Sheet Form

SPECIAL LABOR	CORRECTION FACTOR AREAS
[x] Cartage	Ducts 17ft High
[ ] Shop Drawings	
[x] Field Sketching	
[x] Testing and Balancing	**ALTERNATES**
[ ] Leak Testing	
[x] Service	
[ ] Temporary Heat	
[ ]	**ADDENDUMS**
[ ]	No. 1, Feb. 26, 1987
[ ] Engineering	
[ ] Sleeves	
[ ] Removal	**CONSTRUCTION SCHEDULE**
[ ] Property Protection	12 Months
[ ] Cut Openings and Patch	
SUB CONTRACTORS AND RENTALS	
[ ] Insulation	**BIDDING PLAN AND SPEC NO.'S**
[x] Temperature Control Electric	M1 - M3
[ ] Piping	Section 15
[ ] Electrical	
[x] Kitchen Hood Fire Protection System	**NOT INCLUDED**
[ ] Concrete Pads	Power Wiring
[x] Cranes	
[ ]	
[ ] Testing and Balancing	
[ ] Excavating and Backfilling	
[ ] Cutting and Patching	
[ ]	
END OF BID ITEMS	
[x] Sales Tax	
[x] Permits	
[ ] Bid Performance	
[ ]	
[ ] Travel Pay	
[ ] Room and Board	
[x] Wage Increases	
[x] Material Price Increases	
[ ] Premium Time	
[ ] Contingencies	

Wendes Engineering and Contracting Services

Figure 5-4. Sample Filled Out Bid Checkoff Sheet Form *(Cont'd)*

(Proceed to next page...)

QUOTATION CALL LIST AND PRICE COMPARISON

The quotation call list is an organized written guide for notifying equipment suppliers and subcontractors that you want quotations from them on the project being bid and for making price comparisons after the quotations come in.

QUOTATION CALL LIST AND PRICE COMPARISON

Job Zayre's Discount Store Due Date Feb. 28, 1981

Location 135 Busse Road Estimator

Architect/Engineer Date

ITEM	MANUFACTURER	SUPPLIER	PHONE	QTY	QUOTED PRICE
Roof Tops	McQuay	CTF	338-2681	6	$131 214
	Carrier	—	829-7180		124 978
	Trane	—	567-8492		126 096
Roof Fans	Penn	ARCO	497-8100	2	988
	Cook	Air Products	433-7780		1 014
	Jenn	Filkins	429-8155		1 109
Grilles, Diff	Titus	Air Products	—	27	2 319
	T + B	—	299-8150		2 225
	Agitair	Kreston	276-3181		2 380
Electric Coils	Indeeco	CTF	—	6	—
	Chromolox	Electric Supply	273-5225		—
Temperature Control	Honeywell	—	667-8010		2 400
	Johnson Svc	—	769-5140		2 728
	Robert Shaw	—	415-9761		2 614
Gas Piping	McCarthy		792-4088		5 400
Kitchen Fire Protection System	Ansul		329-8110		900

Wendes Engineering and Contracting Services 7/80 Form EST 104

Figure 5-5. Sample Filled Out Quotation Call List and Price Comparison Form

QUANTITY TAKEOFF AND EXTENSION SHEET

The quantity takeoff sheet is a general form for taking off and listing types, sizes, quantifies, etc. of the various items required in a bid and then for using as a worksheet to extend the material amounts, labor, costs, etc. and total up.

QUANTITY TAKEOFF AND EXTENSION SHEET

Page ______

Date 2/26/81

Job Zayre's Discount Store

DESCRIPTION: DIFFUSERS, GRILLES

Drwg Flr or Sys	Size or Item	Quantity	Labor Ea	Labor Total				
M1	24"ϕ Diffuser	18	2.0	36.0				
	16"ϕ "	2	1.6	3.2				
	12x12 Register	2	.8	1.6				
	24X12 "	4	1.1	4.4				
	48X12 "	1	1.5	1.5				
	48x24 Egg Crate Grilles	20	.2	4.0				
		Total Hours		50.7				

Wendes Engineering and Contracting Services 10/90 Form ESI 105

Figure 5-6. Sample Filled Out Quantity Takeoff and Extension Sheet Form

SUMMARY SHEET

The summary sheet is used as a line item summary of all the different items included in the estimate.

It should be divided into the major divisions of a bid, quoted equipment, ductwork, piping, specialties and accessories, miscellaneous labor, etc.

The total amounts of material quantities, labor, etc. are transferred from duct and piping takeoff sheets, quantity takeoff sheets etc. to this summary sheet.

SUMMARY SHEET

Date Feb. 27, 1981

Page

Job Zayre's Discount Store

		MATERIAL COST		SHOP hr		FIELD hr	
		UNIT	TOTAL	UNIT	TOTAL	UNIT	TOTAL
6	Roof Top Units, 38 Tons ea, 14,700 CFM	$	126,096	—		21	126
	DX, Gas, Economizers	—		—		—	
2	Roof Exhaust Fans, 1000 + 400 CFM		988	—		3	6
2	Prop Fans, Shutters, 1000 CFM ea.		300	—		1½	3
20	Diffusers, (18) 24"ϕ, (2) 16"ϕ		2,225	—			51
7	Registers, (2) 2x12, (4) 24x12, (1) 48x12		—	—		1	2
20	Egg Crate Grilles, 48x24		200	—			10
3	Grease Filters		60	—			1
	(Total equipment $129,869)						
	Galvanized Ductwork, LP, 24,084 lb	$.34	8,189	61 lb/hr	359	31 lb/hr	774
	Lining 1" thick, 1½ lb w/cement, pins 17,000 SF	.32	5,446	70 SF/hr	243	—	
	Black Iron Kit. Duct, 16 Ga. 240 lb	.26	63		6.0		10
3	Turning Vanes 36 lb	.70/lb	25	.5	1.5	—	
6	Splitter Dampers 16 lb	3.00	18	.33	2.0	—	
	Kitchen Exh. Hood, 5'x2'x2' 50 lb	.26	39		10.0		8
	(Total materials $13,780)						
	Cartage, 12 loads 20,200 lb	—		3	48.0	—	
	Field Measuring, 20 pieces	—		—		.5	10
	Testing and Balancing	—		—		—	
	Totals	$	143,649		670 hr		1003 hr

Wendes Engineering and Contracting Services 9/80 Form EST 106

Figure 5-7. Sample Filled Out Summary Sheet Form

TELEPHONE QUOTATIONS

The telephone quotation form is for recording quotations which come over the phone, in an organized, complete and readable fashion.

It includes a checkoff list on the bottom, covering critical aspects of a quote such as, if they meet plans and specification requirements, addenda, taxes, freight, lead times, etc. A box is provided for exceptions on items not included.

Date Feb. 26, 1981

TELEPHONE QUOTATION

Job Zayre's Discount Store	
Supplier Trane	Phone 567-8492
	By Jim Ritter

QTY	MFGR	DESCRIPTION	ACCESS-ORIES	AMOUNT	
				Each	Total
6	Trane	Roof Top Units		$	126,096
		DX, Gas Fired,			
		Curbs, Economizers			
		38 Tons each			
		14,700 CFM each			
		Grand Total			

NOT INCLUDED					
		Thermostats			
		Isolators			

Meets plans and specs ☑	Taxes included ☐ No
Addendums included ☑	Freight included ☑
Type materials correct ☑	Lead time required 10 weeks
	Price good for 90 days

Quote Received By: Fred Schmidt

Wendes Engineering and Contracting Services 9/80 Form No. ESI 107

Figure 5-8. Sample Filled Out Telephone Quotation Form

BID RECAP SHEET

The purpose of the bid recap and markup sheet is to:

1. Recap the direct costs on labor, raw materials, equipment and subcontractors and total them.

2. Put markups on each group and total the overhead markup.

3. Put on a profit markup.

4. Total everything for a bottom-line bidding price.

BID RECAP AND MARKUPS

Job Zayre's Discount Store Due Date Feb. 28, 1981

Location ____ Estimator FS

	HOURS	WAGE RATE	COST	
Shop Labor	670	$21.00	$ 14 070	
Field Labor	1003	$21.00	$ 21 063	
Wage Increase Shop			$	
Wage Increase Field	500	1.50	$ 750	
Overtime			$	
Travel Costs			$	
			$	
			$	
			TOTAL LABOR	$ 35 883
Raw Materials			$ 13 780	
Equipment			$ 129 869	
			$	
Sales Tax			$ 7 182	
			TOTAL MATERIAL AND EQUIPMENT	$ 150 831
Subcontracts Temperature Control			$ 2 400	
Gas Piping			$ 5 400	
Kit Fire Protection System			$ 900	
			$	
			$	
			$	
			TOTAL SUBCONTRACTS	$ 8 700
			TOTAL DIRECT COSTS	$ 195 414
Overhead On Labor		40%	$ 14 353	
Overhead On Material and Equipment:		10%	$ 15 083	
Overhead On Subcontractors:		10%	$ 870	
			TOTAL OVERHEAD	$ 30 306
(16 % of Total Direct Costs) (% of Sales)				
			TOTAL DIRECT AND INDIRECT COSTS	$ 225 720
Profit: 5 % of Total Costs			$ 11 286	
Performance Bond: 1.5 % of Total Bid			$ 3 555	
Financing Costs: Amount $; %			$	
			$	
			TOTAL BID PRICE	$ 240 561

Budget Check: $240,561 ÷ 80,000 sq ft = $3.00/sq ft.
÷ 228 tons = $1055/ ton

Wendes Engineering and Contracting Services 7/80 Form FSI 108

Figure 5-9. Sample Filled Bid Recap Sheet Form

BIDDING RECORD

The purpose of the bidding record form is to have a written record of to whom phone bids were given:

- What was the amount?
- What were the inclusions and exclusions?
- On what plans, specifications and addenda were the bids were based?

BIDDING RECORD

Job Zayre's Discount Store Due Date Feb. 28, 1981

Location ______ Time 2 P.M.

BID SUBMITTED TO:

Company	Name	Phone	Amount	Remarks
Kemper, G.C.	T. Richman	981-2038	$240,560	
Cochran Bldrs	F. Andrew	439-9252	$240,560	
G. Quinn Contr.	Mc Donald	255-0070	$239,660	No fire protection
Bohling Bldg.	R. Burke	870-8662	$240,560	
Sullivan Constr.	H. Kushner	989-6311	$235,160	No gas piping

INCLUSIONS	EXCLUSIONS
HVAC Equipment	Power Wiring
Sheet Metal	Starters
Temperature Control	Painting
Insulation	Framing Roof Openings
Kitchen Hood Fire Protection Sys.	
Gas Piping	

ADDENDUMS	ALTERNATES
No. 1 Feb. 26, 1981	

Drawings Included M1 - M3

Specifications Included Section 15

Remarks ______

Wendes Engineering and Contracting Services 1/81 FORM EST 109

Figure 5-10. Sample Filled Out Bidding Record Form

CALCULATING LABOR COSTS PER HOUR

The following form ensures that all the components of the wage rate are included in the rate used in a bid.

This includes:

- Base wage rate
- Normal union fringe benefits
- Federal and state payroll taxes
- Insurance
- Dues

In the bottom section the contractor is able to determine the cost per pound of ductwork, based on the calculated wage rate above.

CALCULATING LABOR COSTS PER HOUR

Location White County Date Feb. 27, 1981

☑ Union; Local No. 303 Contract Expiration 6/1/81 Non Union ______

FRINGE BENEFITS PER HOUR

	Journeyman	Foreman	General Foreman	Other
Base Rate	$ 15.00	$ 16.50	$	$
Welfare (Medical) (%)	.95	.95		
Pension (%)	1.22	1.22		
Apprentice Fund	.14	.14		
National Training Fund				
Vacation, Savings or Other	.36	.36		
Industry Fund	.18	.18		
TOTAL BENEFITS 19 %	$ 2.85	$ 2.85	$	$
TOTAL WITH BASE	$ 17.85	$ 19.35	$	$

PAYROLL TAXES AND INSURANCE

F.I.C.A. 6.65 %	1.00	1.10		
Workman's Comp. 8.00 %	1.20	1.32		
Federal Unemployment .60 %	.09	.10		
State Unemployment 2.70 %	.41	.45		
Liability Insurance 1.25 %	.19	.21		
Property Insurance 1.00 %	.15	.17		
Association Due %				
TOTAL TAXES & INS. 20.2 %	3.04	3.35		
TOTAL BASE, BENEFITS, TAXES, INS	20.89	22.70		

COST PER POUND BREAKDOWN

Typical Low Pressure Galv 25% Fittings, 24 Ga. Average	LBS/HR	COST/LB	LBS/HR	COST/LB
Material		.33		
Shop Labor	44	.48		
Field labor	25	.84		
Shop Drawings	200	.11		
Cartage	800	.03		
TOTAL DIRECT COSTS		$ 1.80		$
Indirect Overhead 25 %		.45		
TOTAL COSTS		2.25		
Profit 5 %		.11		
TOTAL SELL		$ 2.36		$

Wendes Engineering and Contracting Services 1/81 FORM EST 110

Figure 5-11. Sample Filled Out Calculating Labor Costs Per Hour Form

COMPANY MARKUP CALCULATION SHEET

The company markup calculation form contains easy-to-use formulas for calculating single and dual markups on total direct costs on the projects, and for the total selling cost of labor.

Steps 1 and 2 require determining what the anticipated sales will be for the forthcoming 12 months and what the anticipated breakdown of overhead, labor, material subs and profit will be.

COMPANY MARKUP CALCULATION SHEET

Date JAN. 2, 1985

Company BENSON HEATING + A/C Period ____________

1. Anticipated Sales For Year $ 2,000,000 — Percent of Sales

2. Total Indirect Overhead and Administration Costs for Year $ 350,000 17.5%

   Profit Desired 5 % $ 100,000 5.0%

   Total Anticipated Direct Costs for Year (Material & Labor & Subs) $ 1,550,000 77.5%

   Breakdown: Labor (Includes fringes, payroll taxes, ins.) $ 700,000

   Material and Equipment $ 750,000

   Subs $ 100,000

3. SINGLE MARKUP NEEDED ON TOTAL DIRECT COSTS

   Percent For Overhead Only: Overhead Costs / Direct Costs = $ 350,000 / $ 1,550,000 = 22 %

   Percent For Overhead and Profit: Overhead & Profit / Direct Costs = $ 450,000 / $ 1,550,000 = 29 %

4. SIMPLIFIED DUAL MARKUP FOR OVERHEAD — Amount of Markup For Year

   Markup on Materials and Equipment 10 % x $ 750,000 = $ 75,000

   Markup on Subs 5 % x $ 100,000 = $ 5,000

   Total Overhead on Mat. & Subs $ 80,000

   Percent Markup on Labor for Overhead = (Total Ovhd) - (Matl. & Sub Ovhd) / (Labor Costs) =

   $(350,000) - ($ 80,000) / $(700,000) = $ 270,000 / $ 700,000 = (38.6) %

5. TOTAL SELLING COST OF LABOR PER HR WITH DUAL MARKUP SYSTEM

   Wages per hr (incl. fringes, ins. and taxes) $ 27.50

   (Wages) x (percent overhead markup on labor) $ 10.45

   Profit 5 % (on wages and overhead markup) $ 1.90

   Total $ 39.85

Wendes Engineering and Contracting Services 1/84 FORM EST 117

Figure 5-12. Sample Filled Out Company Markup Calculation Sheet Form

PER PIECE DUCT TAKEOFF SHEET

Estimating ductwork labor by the piece is the most accurate and clearest method available for contractors.

The takeoff involves listing the duct size, type, quantities on fittings, and lengths on straight duct.

The extension of material involves totaling footages per line, entering the weight per running foot and multiplying for the total material weight on each line.

The extension of labor involves totaling the quantity of pieces, looking up and entering labor hours per piece for the shop and field and multiplying out for the totals per line.

After the lines are extended the columns are totaled.

PER PIECE DUCT TAKEOFF SHEET

Job Terrace Restaurant Drawing M-1 System S-1, S-2, E-3, E-4, ☐ Lining

Type Duct: ☑ Galv, ☐ LP ☐ HP, ☐ Other ☐ Insulation

DUCT SIZE	TYPE DUCT*	EQUIVALENT LINEAR FEET PER PIECE	TOT LF	WEIGHT		QTY	SHOP LABOR		FIELD LABOR	
				LBS/LF	Total		Hrs/Pc	Total	Hrs/Pc	Total
48x24	STR	5	5	20.4	102	1	.8	.8	2.3	2.3
48x18	"	5-2	7	18.1	127	2	.8	1.6	2.3	4.6
18x48	SE	2	2	18.1	36	1	2.8	2.8	3.3	3.3
24x18	RE	6-6	12	9.8	118	2	1.5	3.0	1.3	2.6
"	TR	3	3	9.8	29	1	1.0	1.0	1.2	1.2
30x18	RE	7½-7½	15	11.2	168	2	2.4	4.8	2.2	4.4
"	STR	5-5	10	11.2	112	2	.55	1.1	1.6	3.2
"	TR	3	3	11.2	34	1	1.4	1.4	1.7	1.7
30x15	STR	5-3	8	10.7	86	2	.55	1.1	1.6	3.2
"	TR	3	3	10.7	32	1	1.4	1.4	1.7	1.7
30x12	STR	5-3	8	9.8	78	2	.4	.8	1.1	2.2
"	TR	3	3	9.8	29	1	1.2	1.2	1.2	1.2
18x12	STR	5-3	8	7.0	56	2	.3	.6	.9	1.8
16x6	RE	4-4-4-4	16	5.1	82	4	1.3	5.2	1.1	4.4
"	STR	(15)x5'-(5)x5'	100	5.1	510	20	.3	6.0	.9	18.0
"	T	4	4	5.1	20	1	.6	.6	.6	.6
12x6	RE	2-2-2-2-2	10	3.3	33	5	.8	4.0	.6	3.0
"	T	1-1-1-1-1-1-1-1-1-1-1-1	12	3.3	40	12	.4	4.8	.5	6.0
"	STR	(10)x5'-(4)x5'-(3)x5'-(6)x2'	97	3.3	320	23	.2	4.6	.5	11.5
14" φ	T	(7)x1'	7	5.3	37	7	.6	4.2	.6	4.2
10" φ	T	(8)x1'-(4)x1'	12	2.9	35	12	.4	4.8	.5	6.0
36x18	STR	5-5-2	12	15.3	184	3	.55	1.7	1.6	4.8
18x36	SE	2	2	15.3	31	1	2.4	2.4	2.2	2.2
24x12	STR	5-3	8	11.2	90	2	.3	.6	.9	1.8
"	TR	3	3	11.2	34	1	1.0	1.0	1.2	1.2
20x12	STR	5-3	8	7.5	60	2	.3	.6	.9	1.8
"	TR	3	3	7.5	23	1	.8	.8	1.0	1.0
24x6	STR	5-3	8	7.0	56	2	.3	.6	.9	1.8
20x6	SE	5	5	6.1	31	1	1.3	1.3	1.1	1.1
"	STR	5	5	6.1	31	1	.3	.3	.9	.9
12x12	STR	5	5	4.4	22	1	.2	.2	.5	.5
24x24	"	5	5	11.2	56	1	.4	.4	1.1	1.1
12" φ	T	(5)x1'	5	3.3	17	5	.4	2.0	.5	2.5
		Total	413		2,719 LB	125 PCS		63.7 HR		107.8 HR

Mendes Engineering and Contracting Services Form No. EST 120

*STR: Straight; RE: Radius Ell; SE: Square Ell
TR: Transition; T: Tee

Figure 5-13. Sample Filled Out Per Piece Duct Takeoff Sheet Form

PER POUND TAKEOFF SHEET

In the per pound method of takeoff of galvanized ductwork, straight ductwork and fittings are taken off separately.

1. Duct sizes and lengths are taken off and listed. The footage is totaled on each line and the weight per running foot entered.

2. Then the weights are totaled per gauge, totaled for straight and fittings separately and the percentage fittings is determined.

3. The labor factor in terms of pounds per hour is then looked up based on the ratio of fittings and the average gauge.

PER POUND DUCT TAKEOFF SHEET

Job Zayres Drawing M-1 System ______ ☒ Lining 1", 1½ lb

Type Duct: ☒ Galv, ☒ LP ☐ HP, ☐ Other ______ Duct Elevat. 17' ☐ Insulation ______

DUCT SIZE	SQ FT PER LINEAR FEET	LINEAR FEET	TOTAL LINEAR FEET	LBS PER LF with WASTE	0-12 26ga	13-30 24ga	31-54 22ga	55-84 20ga	85 up 18ga	T V	V D	F C	F D
		STRAIGHT DUCT											
56X30		102-30	132	32				4224	lb				
54X20		402	402	21			8442						
30X18		420	420	11.2		4704							
30X14		56-100	156	10.3		1607							
24X12		40-40	80	8.4		672							
32X14		28-100-40	168	13			2184						
12X6		15	15	3.3	50								
12X12		6	6	4.4	26								
			1379'		76	6981	10,626	4224	lb				
		Total Straight			21,907	lb							
		FITTINGS											
56X30		18-18	36	32				1152					
54X20		18	18	21			378						
24"φ		36	36	7.5		270							
30X14		2-	2	10.3		21				2	2		
24X12		6-6-6-6-6-6	36	8.4		371					3		
32X14		1-3	4	13			52			1	1		
16"φ		3	3	5		15							
12X6		3-2	5	3.3	18								
			140'		18	677	330	1152	lb				
		Total Fittings			2,177	lb							
		Total Weight			24,084	lb							
		Percentage Fittings			2,177 / 24,084	= 9%							
		Average Size			22	Gauge							

Wendes Engineering and Contracting Services 9/80 Form No. ESI 122

Figure 5-14. Sample Filled Out Per Pound Duct Takeoff Form

REQUIREMENTS OF A GOOD PIPING ESTIMATOR

A competent and reliable piping estimator must possess the following background knowledge, skills and abilities:

ESTIMATING PRINCIPLES AND PROCEDURES

1. He must follow sound efficient *procedures* for preparing estimates, such as:
   a. Become thoroughly *familiar* with the project, the types of systems and piping, valves etc. involved in the scope of work, before starting a detailed takeoff.
   b. Be familiar with *budget* estimating: HVAC costs for different buildings based on cost per sq. ft. of building or cost per ton of air conditioning; amount of piping per sq. ft. of building or by the average size; cost of piping per linear ft. or per sq. ft. of building.
   c. Know the major *divisions* of an estimate:
      Equipment
      Piping, Valves
      Accessories, Specialties
      Special Labor
      Sub-Contractors
      End of Bid Factors (such as sales tax)
      Markups for Overhead and Profit
   d. Must be familiar with *detailed scope* of what is required in a piping estimate.
   e. *Highlight* drawings before the takeoff
   f. Follow systematic overall *procedure*
      Study the plans and specs
      Send out quotation requests
      Highlight Drawings
      Make Takeoffs and Extensions
      Summarize
      Recap and Markups
   g. Do constant systematic *checking* on each part as he goes along and overall at the end. Double check everything.

2. He must have the ability to *read blue prints,* recognize symbols, types of pipe lines, types of equipment and systems, etc.

Piping Systems

3. He must be knowledgeable of *types of piping systems* such as:
   Low, Med and High Temp *Hot Water* Systems
   Low, Med and High Pressure *Steam* Systems
   Chilled Water Cooling Systems
   Refrigeration Systems
   Hot and *Cold* Water Systems
   Oil and *Gas* piping

   He must not only recognize the various types of systems on plans, but he must know all of the components required in them, whether shown on plans or not.

4. He must know about different types of *piping system configurations* such as:
   1,2,3 and 4 pipe systems
   Reverse and Direct Return
   Constant and Variable Volume
   Closed and Open Systems

Pipe and Fitting Materials

5. A piping estimator must know about different types of *pipe* and *fitting materials,* manufacturing methods, types of fittings in each category and applications to systems.

Black Steel pipe:	Sched 20, 40 and 80, A53, A120, A106, Seam and Seamless Pipe, TC and PE Ends
Black Fittings:	Malleable, Butt weld, Forged and Black Cast Iron Fittings
Copper Tubing:	L.K.M, ACR Soft and Hard Tubing, plus DWV
Copper Fittings:	Wrought and Cast Fittings
Pressure PVC:	Sched 40, 80, Socket and Threaded Pipe and Fittings
Galvanized Pipe:	Sched 40 and 80
Galvanized Fittings:	Malleable
Cast Iron Soil Pipe:	Hub and Spigot, No Hub

6. He must know the *applications* of different types of pipe and fittings materials to various systems:
   Recirculating Water, 250 F
   For 2" dia and under
   Black Steel A53 Seam Sched 40 Threaded Pipe
   Malleable Threaded Fittings

For 2" to 12" dia
Black Steel A53 ERW Welded Pipe
Butt Weld Fittings

Steam and Condensate
For 2" dia and under, 90 lb
Black Steel A53 Seam Sched 40 Threaded Pipe
Cast Iron Threaded Fittings
For 2" to 12" dia, 250 lb
Black Steel Standard Welded Pipe
Wrought Steel Weld

Refrigerant
Copper L,K, ACR Hard Tubing, Brzed
Wrought Copper Fittings

Underground Water
Through 12" dia, 350 lbs
Copper K, Hard Tubing, 95-5 Solder
Wrought Copper Fittings

Potable Water Inside Building, 350 lbs
Copper L Hard Tubing, 95-5 Solder
Wrought Copper Fittings

Fittings and Connections

7. He must be familiar with different types of *fittings* and those available with all the different types of materials:
   Long and Short Radius 90 and 45 degrees *Elbows*
   Straight and Reducing *Tees*
   Concentric and Eccentric *Reducers*
   Straight and Reducing 45 degree *Laterals*
   Caps, Plugs, Unions, Adapters, Couplings
   Weldolets, Threadolets and Sockolets
   Threaded, Slip On, Welded Neck *Flanges*
   Straight and Reducing *Wye* fittings, DWV
   Combination Fittings, DWV
   Bends, 1/8, 1/4 etc., DWV
   Straight, Reducing Tapped *Sanitary Crosses,* DWV
   Traps, Cleanouts, DWV

8. He must have thorough knowledge of the different types of *piping connections.*

   Steel: Threaded, Butt Welded, Flanged, Grooved, Socket Weld
   Copper: 95/5, Solder Brazed
   PVC: Solvent, Heat Fusion, Threaded
   Cast Iron: Soil Pipe, Hub and Spigot, No Hub

9. He must be familiar with various types of *hangers* and *supports* such as rings, cleaves, spring, riser clamps, etc.

Labor

10. He must know *sources of labor* such as MCA and NAPHCC association labor tables, other manuals available, cost records, etc. He must know the *methods* of estimating piping labor such as pipe per ft. and fittings per piece (based on joints), pure per joint labor method or per diameter inch. He must apply *labor multipliers* whenever needed and do so with reasonable accuracy.

11. A piping estimator must be well versed in *piping installations,* the steam fitters plumbers trade, in the operations involved in installations and with tools, scaffolding, etc.

Pricing

12. He must know sources of *pricing* such as piping supply houses, list pricing services such as Trade Services, Harrison, Alpriser, etc. He must be able to use quotations, pricing estimating manuals, etc.

Values and Specialties and Equipment

13. He has to know about *valves:*

    Bronze, Iron, Steel, Plastic, etc.
    Gate, Ball, Globe, Butterfly
    Check, Strainer
    Angle, Diverter
    Steam Traps
    Pressure
    Temperature Control
    Balancing, Gas Cocks
    Refrigeration
    Three Way (Combo check, gate and balancing)

14. He must know about *specialties* such as:
    Air Separators, Air Vents, Bleeders
    Rolatrols (Combo air separator and strainer)
    Pressure Reducing Valves
    Receivers, Sight Glasses, Dryers, Filters
    Vacuum Breakers, Drip Legs, Converters

15. He must know about *HVAC Equipment*
    Pumps: Centrifugal, Inline, Single and Double Suction
    Boilers, Unit Heaters, Baseboard, Heating Coils
    Chillers, Cooling Towers, Cooling Coils
    Compressors, Condensers

16. A piping estimator has to know about *gages* for temperature, pressure and flow readings.

Wage Rates, Unions, Jurisdictions

17. He must know about *wage rates,* fringe benefits, federal, state and local taxes, insurance, etc.

18. He must be knowledgeable about union, trade and local *labor jurisdictions* and he must know about building *codes.*

Design

19. He must have some familiarity with piping system *design* such as typical flow rates, pressures and sizes and with the sizing and selection of equipment.

Other Trades, Types of Buildings

20. He has to be familiar with *other trades* such as piping, insulation, temperature control, electrical and excavation.

21. He must be familiar with all types of *buildings,* commercial, institutional, industrial, their general sizes, layout, etc. and with the sequence of general construction work.

Markups

22. A good piping estimator must be generally familiar with financial statements such as profit loss and balance sheets. He must be able to determine the correct markup for overhead and profit for his company and for the particular job he is bidding.

He should understand how overhead costs are *pro-rated* onto direct material and labor costs for different projects, for different levels of sales and overhead costs, for different ratios of material to labor etc.

Skills, Traits Required

23. Estimating requires a host of skills, mathematical, mechanical, reading, writing, visualizing and drawing. It requires being methodical, analytical, and realistic.

24. It absolutely demands that the estimator be *reliable,* that he be thorough in his understanding of the project, of its scope, in takeoffs, interpretations, extensions, summaries, recaps.

Thus, the knowledgeable, proficient and reliable estimator as described above will be able to produce complete and accurate estimates, which in turn become the required foundation blocks of successful contracting.

PIPING TAKEOFF SHEET

Pipe and fittings are listed separately

1. The diameter, type item, footage for pipe and quantities for fittings, are taken off of blueprints and listed.

2. Footages and quantities are totaled per line, labor and price factors looked up, entered and then extended.

3. The cost and labor columns are then totaled.

PIPING TAKEOFF SHEET

Job Atrium Restaurant Drawing M-1 System Gas Factors ____________

Type Pipe B.I. Connections Screwed ☐ Insulation ____________

DIAMETER	ITEM	QUANTITY OR LINEAR FEET	TOTAL	MATERIAL COST		ERECTION LABOR		
				Unit	Total	Equiv. Joints	Unit	Total
2" φ	Pipe	2-10 (Feet)	12	2.49	30.00	2	.64	1.28
1½" φ	"	23	23	1.78	41.00	2	.44	.88
3/4" φ	"	2-6-2-9-6-2-4-6						
		2-8-2-8-2-8-2-8	79	.86	68.00	17	.28	4.76
1" φ	"	30	30	1.18	35.40	3	.34	1.02
		(Total Pipe)	144'					
2" φ	90° Ell	11 (Pieces)	2	3.50	7.00	4	.64	7.04
3/4" φ	"	卌 卌 IIII	14	.60	16.80	28	.28	7.84
2x2x1	Tee	1	1	—	5.05	3	.64	1.92
1½x1½x¾	"	1	1	—	4.63	3	.44	1.32
3/4" φ	Unions	卌 III	8	.96	7.68	16	.28	4.48
2" φ	"	1	1	—	5.90	2	.64	1.28
1" φ	Couplings	1	1	—	1.35		—	—
1 ' φ	"	1	1	—	2.20		—	—
		(Total Fittings)	29					
3/4" φ	Gas Stops	卌 II	7	4.60	32.20	2	.28	.56
	Hangers		15	1.26	18.90		.5	7.50
		Totals			$ 276.11			40 hr

Wendes Engineering and Contracting Services — Form No. ESI 130

Figure 5-15. Sample Filled Out Piping Takeoff Sheet Form

COMPUTERIZED ESTIMATING

WenDuct and WenPipe Systems

Fast Accurate, Easy to Use

Benefits of Computerized Estimating

Bid two to three times as fast with the WenDuct and WenPipe systems. Prepare more accurate and complete bids. Save time and money.

Easy to use, menu driven, self prompting system with a short learning curve that guides you through the natural flow of estimating.

Complete *WenDuct* Sheet Metal Estimating System Covers:

- All types of HVAC *SYSTEMS*
- Rectangular and Round *HVAC GALVANIZED* Ductwork
  Low, Medium and High Pressure
- All types of *CONNECTIONS*
  Cleats, Ductmate, TDC etc.
- *SPIRAL* Pipe and Fittings, Single
  and Double Skin, Round and Oval
- *FLEXIBLE TUBING*
- *FIBER GLASS* Ductboard
- *HEAVY GAUGE* Metals
  Up to one half inch thick
  Black Iron, S.S., Aluminum, etc.
- All Types of *DUCT ACCESSORIES*
- All Types of *EQUIPMENT*
- *LINER* and *DUCT WRAP*

The WenDuct estimating software package covers everything needed to produce COMPLETE sheet metal estimates.

Complete *WenPipe* Estimating Systems Covers:

- All Types of Piping *SYSTEMS*
  HVAC: HWH, CHW, Refrigeration, Steam, Gas, Oil
  PLUMBING: Soil Pipe, Drainage, Storm, Vent, Hot
  and Cold Water

FIRE PROTECTION
PROCESS PIPING

- All Types of Pipe *MATERIALS*
  Black Steel, Copper, PVC

WenPipe Estimating Systems Covers (Continued):

Galvanized, Stainless Steel Cast
Iron Soil Pipe, Ductile
Poly Pipes, FRP, etc.

- All Types of *FITTINGS,* HVAC, DWV
  Malleable, Butt Weld, Flanged, Grooved, Wrought Copper etc.
- All Types of *CONNECTIONS*
  Threaded, Welded, Flanged, Grooved, Soldered, Brazed, Solvent
  Bell and Spigot, No Hub, Mechanical
- All Types of *VALVES* and *SPECIALTIES*
- All Types of Plumbing *FIXTURES*
- *HANGERS,* Gaskets, Flanges, Couplings
- *TRENCHING* and *INSULATION*
- *EQUIPMENT,* Gages

The Super-Pipe HVAC, Plumbing and Process Piping estimating systems covers everything needed for complete mechanical estimates.

High Speed, Easy To Use Features

- High speed takeoffs with *DIGITIZER BOARD* and PEN or *COMPUTER KEYBOARD*
- *LABOR* and *PRICING TABLES* preloaded, ready to use, user adjustable.
- Automatic *DUPLICATIONS* and *ASSEMBLIES*

ESTIMATING PROCEDURE DIAGRAM
FOR PIPING AND SHEET METAL WORK
Wendes Super-Duct and Super-Pipe Systems
Clear, Direct Process

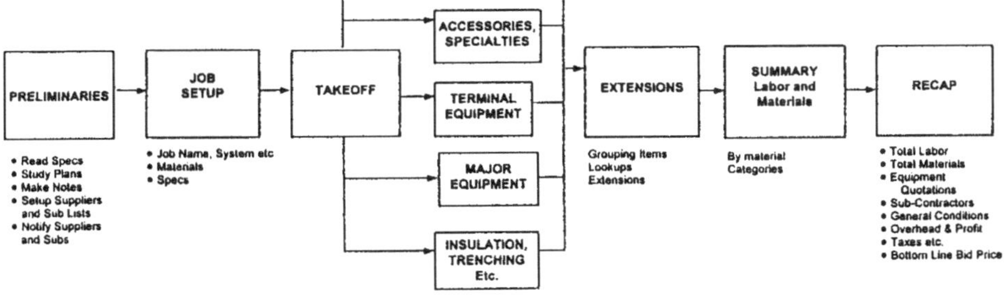

- Automatic generation of *CONNECTION, HANGERS,* and *COUPLINGS.*
- *CALCULATES* all material and labor automatically
- Automatic *PRICE COMPARISONS*
- *AUTOMATIC PRICING SERVICE* Available

Pricing and Labor Tables Preloaded

All pricing and labor figures are preloaded and WenDuct and WenPipe are ready to use out of the box. All tables are user adjustable.

Optional Computer Keyboard Or Digitizer Entry

Since the digitizer template and the computer keyboard are 100 percent compatible you have the option of using either, and switching back and forth as needed. The system is completely portable. You can estimate at home or at other offices.

Complete Estimating Reports

Super-Duct and Super-Pipe automatically looks up, extends and summarizes all *LABOR* and *MATERIAL* needed for complete bids. Clear and easy-to-read reports are printed out. Eliminate all messy, time consuming, error prone, and hand extension summaries.

- SEQUENTIAL TAKEOFF LISTS
- EXTENSION and SUMMARY Sheets
- BILL OF MATERIALS
- EQUIPMENT QUOTATIONS
- SUBCONTRACTOR QUOTATIONS
- Percentage MARKUPS for Overhead and Profit
- Bottom line BIDDING PRICE
- "WHAT IF" PRICE COMPARISONS

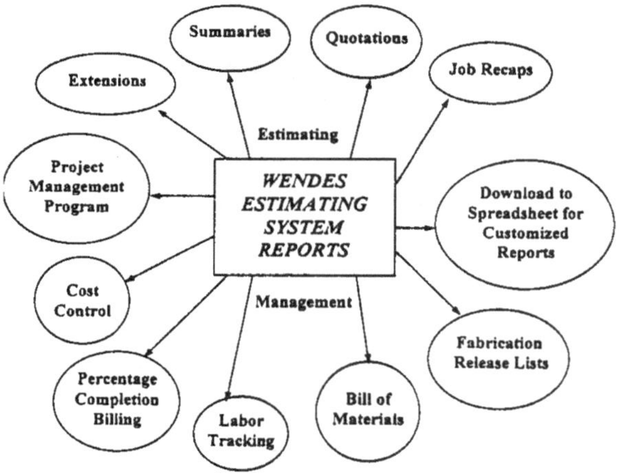

Complete Management Reports

Super-Duct and Super-Pipe can be used as a management tool for more organized and efficient operations.

- MATERIAL LISTS for QUOTATIONS, PURCHASING, FABRICATION AND RELEASES
- SCHEDULING

- AUTOMATIC PRICE COMPARISONS
- LABOR TRACKING
- COST CONTROL and JOB STATUS

CAD Drawing System Downloads Into Estimating System

Can download CAD ductwork drawings from InteliCAD into estimating systems for automatic generation of estimate saving without a takeoff.

Customize Personal Totals Reports

WenDuct and WenPipe estimating system users can now quickly and easily download the standard estimating reports into popular spreadsheets and create their own reports in any manner.

pcANYWHERE™ Modem Service

Perplexing and time consuming service calls can now be greatly simplified, and time on the phone reduced with the new pcANYWHERE™ modem service offered by WENDES systems.

Support, Service and Training Included

Complete support, service, unlimited phone time and training are included in the base price. This covers an instruction manual and free updates, for the first year.

For more information on the WenDuct or WenPipe estimating system, check internet, *www.wendes.com.*

Chapter 6

Hitting Home Runs in Technical Management

The Principles, Procedures and Forms of Home Run Problem Solving, Troubleshooting and Decision Making In Building Engineering and Contracting

Building engineers and contractors are essentially in a problem solving, troubleshooting and decision-making business.

Problem solving is not only an integral part of building system management and of contracting, but also is one of their most common activities. It is a continual succession of meeting objectives, and of overcoming obstacles and difficulties.

One of the biggest factors in the success of building engineers and contractors is how many home runs they hit in the problem solving process. The success of their careers is based on how well they handle the never-ending parade of new problems and decisions confronting them.

Unfortunately, many of the people involved in this area of work are still in the minor leagues when it comes to how well they score at solving their problems. For example, cardboard is used to blank off a combustion air louver because there is a draft when you walk by. Causes are not determined accurately or thoroughly enough on malfunctioning equipment, and the wrong part is replaced. Trial and error replaces sound troubleshooting, and so on.

However, their batting averages can be greatly improved and their field errors greatly reduced.

This chapter is about moving up from the minor leagues to the major leagues. It is about thorough, systematic and accurate problem solving, troubleshooting and decision making. Excerpts are taken from the forthcoming book *Hitting Home Runs in Technical Management*, by Herb Wendes, P.E.

FACED WITH A MULTITUDE OF PROBLEMS

Every day, operation and maintenance engineers, and contractors, are faced with a multitude of problems and decisions, which come in all sizes, varieties and complexities.

- *Operation and maintenance engineers* are confronted with system and equipment breakdowns, service and maintenance. They are faced with making expeditious troubleshooting and effective decisions on what to do and how best to do it.

- They are faced with the proper *operation and performance* of systems.

- They are responsible for effective *preventive maintenance.*

- They are faced with maintaining *comfort conditions* for the occupants, resolving the problems of hot and cold areas, drafty and stagnant areas, negative and positive building pressures, system imbalances, etc.

- They face *demands and conflicts* with occupants, owners, personnel, and architect/engineers.

- They are confronted with *purchase decisions,* costs, budgets, performance factors, timely deliveries, quality, durability, and so on.

- They are confronted with *"repair or replace evaluation decisions"* and life cycle cost analysis.

- They must maintain *indoor air quality.*

- They must meet *building codes,* union requirements, health and safety conditions and legal issues.

Contractors are involved with a continuous multitude of problems and decisions in a similar way:

- They are involved with *purchasing, scheduling, and installation* of equipment, piping, ductwork, etc. in a cost effective and timely manner per the contract documents.

- They are involved with preparing accurate *estimates* and meeting bidding deadlines. They must make sales and negotiate contracts.

- They must manage their *financing*—accounting, cash flow, billing, receiving payments, and overhead costs.

- They are faced with *personnel management*—planning and scheduling work, union requirements, getting and keeping skilled competent labor, doing acceptable quality work, and acceptable productivity.

- They must *coordinate and resolve conflicts* with the architect, engineer, owners, and other contractors in a mutually satisfactory manner.

- They must maintain *cost control* of material, labor and overhead. Inventories must be maintained.

- They must meet *building codes*, union requirements, health, safety and legal issues.

- In the end they must meet the problems of making the systems *perform* per contractual requirements and to the satisfaction of the occupants.

Causes of Problems in Building Systems

Problems with mechanical and electrical systems in buildings stem from the multitude of *components* in the systems, ranging from tiny transistors to huge pumps. They stem from the *complexities* of the systems.

Problems stem from the inevitability of *human error* in designs, installation, equipment, controls, electrical items, balancing systems, changes, maintenance, operations, etc.

Problems are generated from *wear and tear* on the system components.

Problems stem from the uncontrollable *variety of environmental conditions* inside and outside the buildings, and an endless host of other causes.

Problems stem from changes from the original design, remodeling, and changes in the occupants needs and use of spaces.

Move up from the Minor Leagues to the Majors

- Achieve *effective operations and maintenance* of the building systems.

- Satisfy the requirements of the *occupants* of the building, owners and investors.

- Avoid or minimize unnecessary *breakdowns* and down times of systems and equipment.

- Maximize preventive maintenance.
- Improve the monitoring of the systems.
- Improve the handling of deviations in performance.
- Control operational and overhead costs.
- Control *capital expenditures.*
- Help ensure that *occupants* will stay in the building and renew their leases.

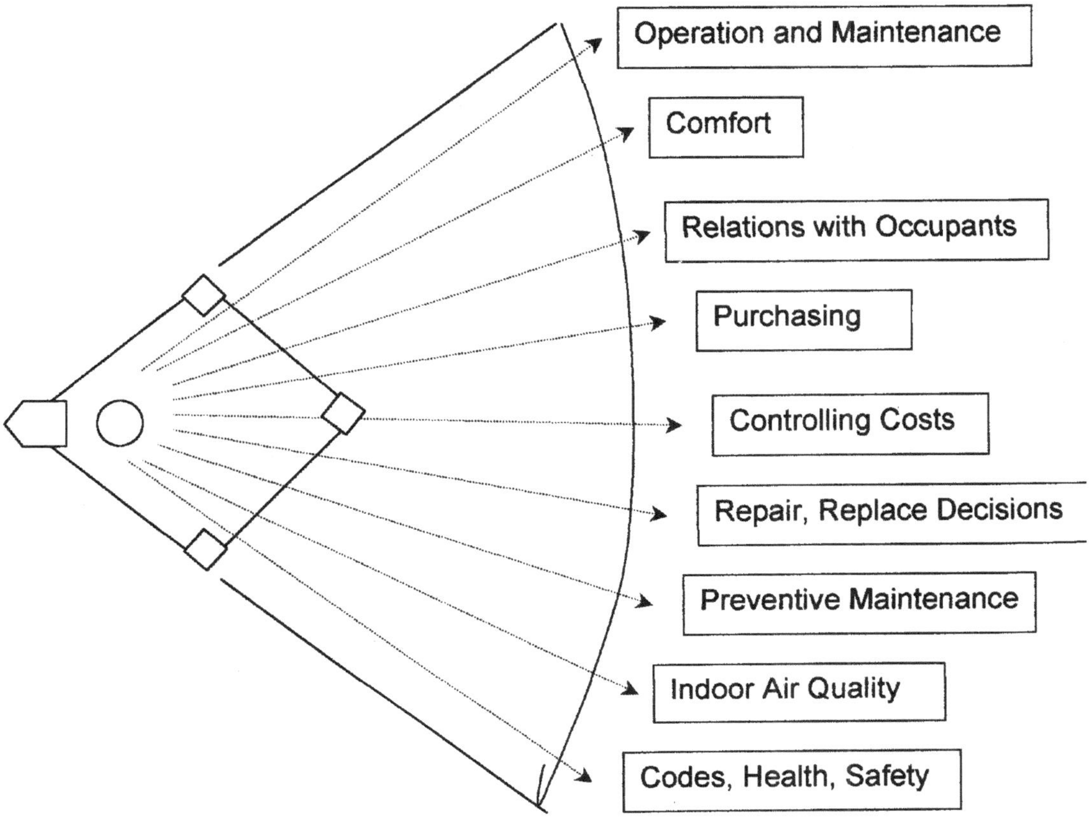

Figure 6-1. Hitting home runs in building management and maintenance.

Benefits of Developing Home Run Problem Solving Skills

The more *knowledgeable* you are about the principles and procedures of hitting home runs in the problem solving and decision making processes, the more conscious you are of where you are and what you are doing in the process and the more deliberately you apply the process, the better you will swing your bat to get home runs or extra base hits, leading to the best solutions and most effective decisions.

- The principles, procedures and forms in this chapter will help *guide you* through the process of greatly improving your ability and performance at home run problem solving and decision-making.
- You will make more *informed, intelligent decisions,* solve problems faster and better, and solve them more effectively, objectively, thoroughly, and systematically. You will perform with *less strife* through the application of the principles and procedures. You will feel great *confidence* when decisions are made through this dynamic process—instead of that sunken feeling in your stomach when uncertain what the results will be.
- You will handle the *troubleshooting* of problems, making choices faster and better. You will avoid *sloppy and fragmentary* problem solving and decision-making, and be able to setup, analyze and evaluate *complex problems* better.
- You will control the *negative, adverse emotions,* which often thwart you from achieving the objectives.
- You will liberate and utilize *imaginations and creativity* powers much more extensively—greatly improve the *logic, thinking, and systematic* handling in problem solving and decision making— increase *alertness* and use senses more effectively.
- You will unravel and manage the *conflict of interest* issues that often plague people in the administration of their duties, and you will become more skilled at conflict resolution.

You will *develop the ability* to hit more home runs, and avoid strikeouts, groundouts, pop outs, and errors. This chapter provides a *proven system* of successfully playing top-notch major league baseball in home run problem solving, troubleshooting and decision-making.

Understanding What Constitutes a Problem

A problem, in its simplest and broadest terms, is something not the way you or your group want it, along with a desire to change it. *The boiler breaks down in your building on*

Christmas day during a frigid cold wave. This is obviously not the way you want it and you desire to change the situation as quickly as possible before it turns into a disaster.

It's the difference between the way things are and the way you or the group wants them. *Rain, mud, and cold weather during construction of a fast track production plant sets the schedule back months.*

Problems are undesirable circumstances and conditions caused by prior events. *An extended heat wave in middle of summer during a peak period causes an electrical power blackout.*

A problem consists of the deviation from a norm, standard or criteria. *In the high school sample problem farther on in this chapter, a 32-decibel sound level, or less, was required for the library, and the actual was closer to 60 decibels.*

The problem may be preventive. *Changing oil in car every three months greatly reduces premature and unpredictable breakdowns and wear on the engine. Cut down on fatty food to prevent conditions, which could lead to clogged arteries and a heart attack.*

A problem may be determining a hidden, elusive cause. *The High School Vibration problem farther on in this chapter illustrates a hidden, elusive cause situation.*

A problem might be finding and selecting the best chiller *to purchase* to replace an old broken one.

A problem may be how to achieve a goal. This involves determining what the gap is between the present circumstances and your goal, and how to achieve the goal. For example your goal may be to keep your car in good shape for 100,000. In the past you have only been able to keep your car in good condition for 50 or 60 thousand miles.

MULTIPLICITY THINKING; AVOID PARTIAL VIEWS

Keep mind open to multiple possibilities. A major principle of home run problem solving is multiplicity thinking—listing and considering a full range of relevant factors in each category of the process.

Think in terms of a full array of possible factors in each category—causes, objectives, options, effects—rather than skimping with just one or two items. *In the high school vibration problem there are nine possible causes listed, nine objectives and also eight possible options.*

Partial views of problems, causes, etc. can be disastrous. *A more limited consideration of the possible causes in the high school vibration problem, and the resultant action based on selecting the wrong cause, would have wasted tens of thousands of dollars, caused an unbearable delay and not resolved the problem.*

Considering the full range of possibilities makes it easier to hold back on conclusions and decisions until sufficient preparation is completed, as well as avoiding overlooking critical factors.

Practice multiplicity. Do not strap your self down with singular items in the problem solving categories of the process.

Home run problem solving involves the ability to gather the full range of relevant factors in each segment of the process. It involves the ability to remember, focus on, and effectively analyze and evaluate these factors.

These techniques of multiplicity thinking, making notes where needed, and prudently holding back on conclusions, are powerful techniques leading to the best generation and selection of a solution.

BE ACUTELY AWARE OF THE PRINCIPLES AND PROCEDURES

- *Be aware of what you are doing and recognize what component you are working on,* in the home run problem solving and decision making process as you journey through it. Be conscious of and focus on what phase you are working on in the procedure... defining the problem, analyzing the cause, establishing objectives, generating solutions, choosing one, implementation... and so on. Being able to focus on a category of the process to achieve an effective result is a major principle of home run problem solving. You will be far more efficient, effective, confident and self-correcting in the process.

- Learn to distinguish and identify the *elements* of a problem and the causes. In the high school vibration problem, the problem elements were the fans, rpm's, motors, isolators, structural steel, concrete floors and so on.

- Deliberately apply the major *principles* of dynamic problem solving and decision-making, such as considering all possible causes, seeking alternate solutions, holding back on conclusions and choices until adequately prepared, generating and listing multiple factors in the various categories of the process.

- Deliberately apply the major *skills* in the home run process, such as the various methods of getting information, productive imagination and creativity, sound reasoning, and being open-minded.

- Dynamic problem solving and effective decision-making is a *complete process* from the time the problem occurs through the final enactment of the successful resolution. It is the entire process from inception to achievement.

- Follow the same process of resolution with *sub-problems as with main problems.* Sub-problems will normally involve a definition of the sub-problem, and some or all of the other components in the process—cause, objectives, options, evaluation, decision-making and implementation. *In the high school vibration problem one of the sub-problems is how to test out if the vibration isolators for the fans are doing the job they are designed to, and if they are part of the cause or not.*

- *Don't necessarily follow a rigid sequence* or step-by-step procedure in the home run problem solving and decision-making process. Rather, follow a more general, multi-directional type thought process, dependent on how your mind responds to the exposure to the information, elements and relationships, of the problem. "You learn the skills and combine them to play the game as circumstances dictate."

- *On the other hand, don't jump around haphazardly.* Once working with a category of a problem, such as the problem definition, objectives or options, generally stick with it until getting something substantial done. Then jump onto other categories of problem.

- *Follow practical more easygoing approach.* Dynamic problem solving and decision making does not necessarily require that you function like a computer, or that you apply complex and sophisticated techniques.

Hitting home runs in problem solving and decision-making is a personal and concrete mental process. It is your brain, memory bank, senses, reasoning, and the accumulation and processing of information in the mind that guides you successfully through the process, not ultra-sophisticated mathematical methods.

OUTLINE OF PRINCIPLES AND PROCEDURES OF HOME RUN PROBLEM SOLVING

The most important principles and procedures of thorough, systematic, accurate and complete problem solving and decision making covered in this chapter are as follows:

Problem, Cause, Inter-relationships

- Become familiar with the *problem.* Gather information. Identify, analyze and define the problem.

- Examine possible causes and suspend judgement until verified. Troubleshoot. Diagnose.

- Identify and understand relevant *problem elements and inter-relationships.*

Objectives and Options

- *Thoroughly establish and evaluate objectives* and conditions that the solution must meet.
- Generate and evaluate *optional solutions.*
- *Hold back* on the choice of an option until sufficient preparatory work on the problem has been completed.
- *Pause* at various times in the process to allow the subconscious mind to clarify items, make connections, generate ideas and solutions, be creative, and allow for mental digestion.

Evaluation of Possible Solutions

- Determine *good and bad effects* of optional solutions.
- Determine *risks.* Evaluate the risk-to-reward ratio as required.
- Determine *odds* of success or failure of options.
- Confirm relevant *assumptions,* relationships, information, generalities, theories, etc. Scrutinize fixed beliefs, biases, and prejudices.

Comparing Options Against Master List of Objectives and Making Decision

- *Compare and evaluate* optional solutions against master list of prioritized objectives.
- Rate the various optional solutions. Follow process of elimination.
- Select the solution that best *satisfies* the prioritized objectives and conditions, and best eliminates the causes. Make decision.

Implementation

- Determine how to *best implement* chosen solution.
- Identify and resolve *implementation* problems.
- Plan and schedule. Take action in an effective and timely manner.
- Monitor, follow up and modify as required.

The foundation of home run problem solving in technical management is high performance problem identification and analysis, accurate cause diagnosis, prioritized objectives of solution, effective generation of optional solutions, sound decision making, and productive implementation.

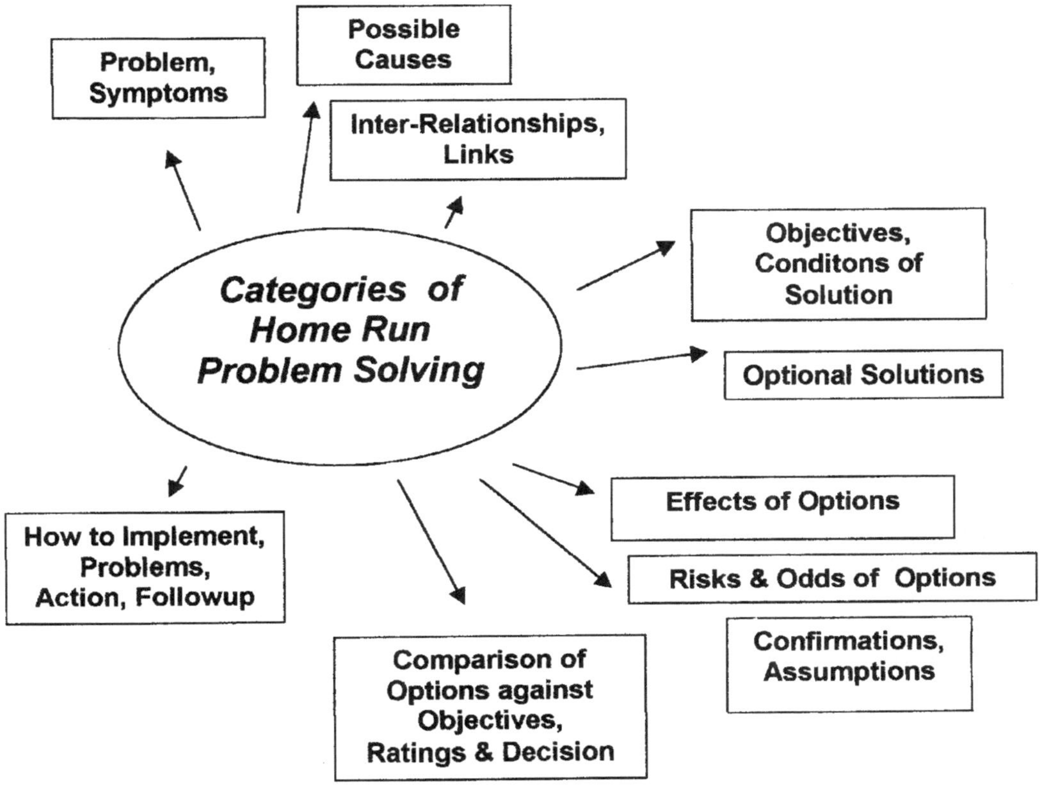

Figure 6-2. These are the major components of home run problem solving and decision making with which a home run hitter is involved.

VIBRATIONS IN HIGH SCHOOL BUILDING

Sample Case History

Here is an actual case history of a problem in a newly constructed high school in the Chicago area illustrating the application of the principles and procedures of home run problem solving and decision making, which I was involved in some years ago. Sample filled in forms follow along with more detailed explanations of the hitting home runs principles and procedures.

Problem

The heating and air conditioning equipment in the mechanical room of the newly constructed high school vibrated, rumbled and was very noisy. The library, which was adjacent to the mechanical room, was intolerable. Tremors shook the bookcases, tables, and chairs, as if there were a mild earthquake. Blinds, floors and ceilings vibrated. The lockers in the locker rooms underneath the mechanical room shook and rattled. The school personnel and school board were horrified and in a rage.

The vibrating, rattling, and throbbing were really resulting symptoms and effects of some root cause. The problem was to search out the root cause.

Information and Research

The mechanical room had almost 30 large pieces of rotating equipment, fans, compressors, chillers, pumps, etc. sitting on extremely heavy concrete inertia blocks with spring isolators. All of this live weight was sitting on a lightweight three-inch-thick concrete floor and metal pan.

Possible Causes

It was finger-pointing time by all parties—the engineer, architect, air conditioning contractor, piping contractor, vibration isolator manufacturer, fan supplier, balancing contractor, and sound consultant. Initially, all were either defensive or blamed others. No one wanted to get stuck with the blame and responsibility of having to stand the cost of rectifying the problem. The architect and engineer were protecting their reputation and blaming the HVAC contractors.

However, despite the protective stances taken, inspections of the building and equipment were made, ideas on potential causes brainstormed, and a list of possible causes was made as follows:

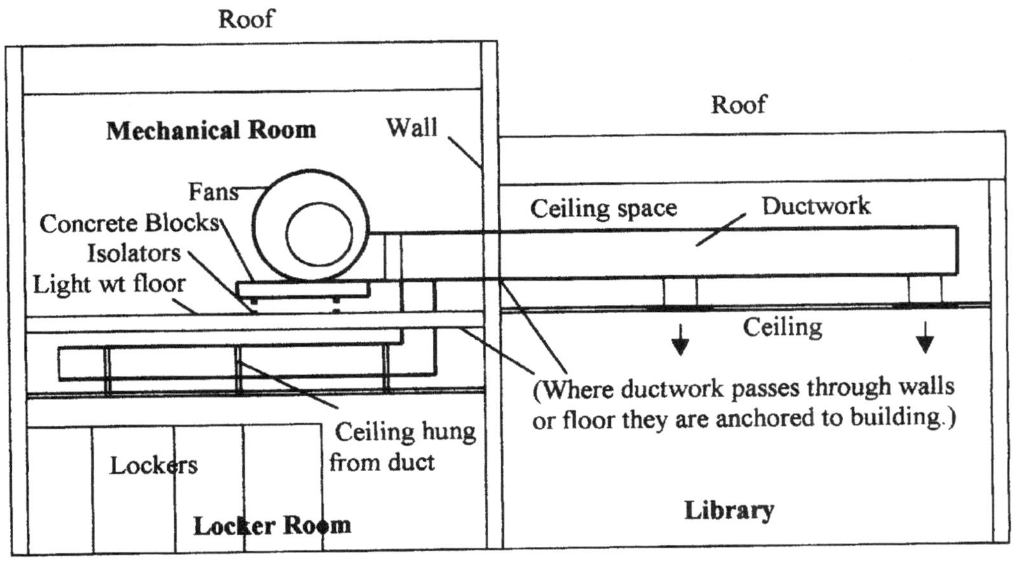

Figure 6-3. Elevation cross section of mechanical room with 30 pieces of rotating equipment transmitting vibrations and loud rumbling noises to the floor of the equipment room, and to adjacent areas of the library and locker rooms.

1. Possible out of *balance* fans.
2. Incorrect *vibration isolators* used under fans.
3. Throbbing *ducts passing through floors and partitions* connected directly to building.
4. *Locker room ceiling* hung from vibrating high velocity ductwork.
5. *Air pulsations* in fan housings.
6. Throbbing ducts pressed against *joists and beams.*
7. Incorrect *acoustical lining* in ductwork.
8. *Debris wedged* under concrete inertia blocks, which equipment was set on, transmitting vibrations to building.
9. Thin, 3" thick, *lightweight concrete floor* in equipment room; insufficient mass to absorb vibrations from such a large equipment load.

Objectives

There were objectives common to all parties involved, personal interest objectives of individual parties, and some conflict-of-interest issues, which were not necessarily stated or publicized. Their collective objective:

The overall goal was to *rid the building of the vibrations* and loud rumbling as quickly as possible, achieving at a minimum short-term acceptable results without total disruptions of the school operations. Further objectives were:

1. The school faculty and staff, the school board, and the architect and design engineer weren't interested in whose fault it was initially. They just wanted *immediate action,* no matter who was responsible or who paid for the rectification.

2. The mechanical contractors wanted to *determine the true cause first,* and not waste time and money on trial and error solutions based on assumptions or self-protection, which might not work in the end.

3. The mechanical contractors involved in doing the work, wanted to know *who pays for the work* they will do testing and rectifying the problem, before they do it, if it is deemed not their responsibility in the end.

4. The school wanted it corrected without *closing the school down.*

5. The school wanted it corrected *without turning heating/cooling systems* off while school was occupied.

6. The school did not want to shut down the effected areas if possible.

7. The corrective measures could not violate the parameters of *comfort, health or building code requirements,* etc.

8. The corrective measures in the end should be in an acceptable realm of the *specifications, contracts, and performance requirements.*

9. The long-term solution must be a *permanent* one.

Options

A list of optional solutions was generated based on the possible causes; however, they were to be selectively tested out first, to determine if they would really solve the vibration problem, before implementing them overall.

1. Balance select fans
2. Replace vibration isolators under fans.
3. Reduce air quantities from fans by lowering Rpm's.
4. Isolate ducts through floor and partitions from building.
5. Install special sound baffles in fan housings.
6. Install special airflow baffles in fan housings to get rid of pulsations.
7. Clad ducts with special lead and sponge rubber.
8. Clean out construction rubble under concrete inertia blocks.
9. Any or all of the above.

Testing

Tests were made to establish the cause of the vibrations and noise by a process of elimination. A priority sequence was established in order of the likelihood of being the actual cause.

1. Select fans were rebalanced, vibration isolators checked.
2. Ducts detached from direct contact with building.
3. Debris cleaned from underneath the equipment inertia blocks.
4. Acoustical lining material in ducts tested for sound absorption.
5. Heavy lead sound baffles put in a fan housing.
6. Airflow baffles put in a fan housing to counter pulsations and rumbling in the housing and in the ductwork.

And so on down the list of possible causes.

None of these tests had any effect on the vibrations and rumbling, and all the cause and effect theories up to that point in the problem solving process, were thrown out. The fans were in balance, isolators were correct per specifications, etc.

Further Information, Research, Consulting Needed

Until this time the architect and engineer avoided any talk about the lightweight concrete floor used for the mechanical room floor, and the fact that the mechanical contractors questioned this during the course of construction. The thought of the floor being the reason leading to the this violent vibration problem was irreconcilable in their minds.

However, the superintendent of the high school district, not influenced by the fears of the architect and engineer and their reputations, called a special meeting in his offices. A representative of the HVAC contractor, a school board member, and a mechanical engineer who happened to be an expert on building vibrations were requested to attend.

The school board member came to the meeting with an instrument for measuring vibration frequencies and amplitudes of solid bodies such as concrete floors and equipment. He recommended that the vibrations of the mechanical room floor and the 30 pieces of equipment be measured and compared.

True Cause and Effect Relationship

Thus, the three went to the high school and measured the vibrating frequencies and deflection of the concrete mechanical room floor and the fans. The expert checked to see if any equipment was vibrating or running at an rpm in the same range or multiple of harmonic frequencies as the floor. He found several large fans were at close harmonic frequencies with the concrete floor.

The relationship of simultaneous harmonic frequencies between different bodies in contact with each other produces a massive combined geometric increase in force. This is a principle of physics. This was vaguely suspected and was in the back of many people's minds, but no one was versed enough to bring it out and test for it.

Hence, a combined harmonic frequency effect of the mechanical equipment room floor and the equipment in it, can be a major problem, because it can set up an amplified deflection in the floor and be transmitted to adjacent connecting parts of the building.

This is similar to the classical example of the bridge collapsing when the soldiers marched across it with a cadence that was a harmonic frequency of the bridge.

Further Testing

After further investigation of the situation it was found that some of the fans were in the harmonic frequency range of the floor. As a test, the two fans with a frequency closest to that of the floor, were *turned off* to see what would happen. Ninety percent (90%) of the vibrations and rumbling stopped, thus confirming the true cause, a simultaneous harmonic frequency situation!

Decisions and Implementation

It was finally decided that the best solution was to lower the operating speeds of these two fans about 15% to 20% in order to get them off the harmonic frequency. This was still in a range of tolerance in regard to ventilation code and heating cooling requirements meeting those absolute objectives. New drive packages were installed on the fans accordingly.

The fans were turned at the lower operating speeds and *the vibrations and rumbling virtually disappeared*. A home one was hit and this proved to be an acceptable solution to all. All readers of this case history who suspected the cause to be a harmonic frequency one—give yourself an "A" in home run problem solving.

Points to Note:

Many of the principles and procedures of home run problem solving, troubleshooting and decision making were employed in this case history:

1. Extensive information was gathered and a number of tests were conducted regarding the problem description, the possible causes and possible solutions.
2. Extensive notes were made during the course of the problem solving process covering the description of the problem, symptoms, causes, tests, optional solutions, etc. These notes promoted easier focus and evaluation of the various factors in the problem, and better communications and interaction between the parties involved.
3. The cause in this actual problem was hidden and elusive to most of the parties involved. The self-interest of the architects, engineers, contractors, owner, etc. was defensive and blaming. A lack of expertise in this particular sphere of knowledge clouded everyone's thinking.
4. A process of elimination of the possible causes was employed. Finally an expert, who had experience with this type of problem, was consulted and this led quickly to the resolution.

BECOME THOROUGHLY FAMILIAR WITH THE PROBLEM, DEFINE ACCURATELY

If you don't know what the problem is, if you are guessing or are assuming what it is, if you are skimming over it and not becoming adequately appraised of it, if you don't know what the deviation is—it means that you really don't know what is wrong. As a result you don't have a sound direction to go in, nor really know what is the right action to take to resolve the problem.

Getting to know the problem means gathering enough information on it, analyzing it, and defining it sufficiently to insure that your understanding is complete and accurate. Know what the problem is, fully and explicitly, before making a decision and taking action.

1. *Become familiar* with the problem—the overall problem and any sub problems involved.
2. *Gather information*. Make observations, examinations, and tests as required. Make notes, sketches, etc. as required. Plan and schedule this work, both long and short term if required.
3. Assess the *importance, urgency and size* of the problem. Decide if the problem will be *confronted* or not, and whether it is a short-, immediate- or long-term situation.
4. Make a *brief problem statement* for a convenient identification and reference, and a broader, more detailed overall statement.
5. Note problem *symptoms*, syndromes, indicators, conditions, and events that occurred prior to the occurrence of the problem,
6. Note the *general and specific aspects* of the problem as required in terms of *what, where, when and extent (www.e)*. Go from general aspects of the problem to more specific ones.
7. Determine, define, clarify, and state what the *deviation* is, if not obvious and it is required. Become familiar with the *standards, norms*, needs, desires, etc. of the particular problem situation.
8. Determine what the problem *is as well as what it is not*. Note what has *changed and not changed.*
9. Note *cause and effect relationships* and what elements are involved in the problem. Determine if cause diagnosis is needed or not.
10. *Consult with others* such as those who are involved or effected by the problem in some way, or experts in the area, as needed.
11. Be aware that there can be *different stages* in a problem situation, that is, a series of cause and effect events, and that you are usually focusing on a certain stage, which you are defining as the problem. Be aware whether you are just *treating the symptoms* of the problem, some intermediate state in the sequence of events, or the actual root cause.
12. Conclude with a full valid definition of the problem.

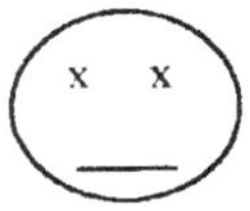

PROBLEM WORK SHEET

Hitting Home Runs by Wendes

Date: *September 5, 1996*

Page ________of ________

PROBLEM.......(brief statement):
Vibrations in high school building.

OVERALL PROBLEM:
The equipment in the mechanical room vibrates and rumbles, effecting adjacent areas so badly that these areas cannot be occupied.

General Area or Category	PROBLEMS (main problems and sub-problems)		Possible Key Causes or Options
	General Description	**Specifics**	
	Extreme vibrations and rumbling in the equipment room being transmitted to adjacent school areas.	*Library and locker rooms*	*Mechanical equipment. Light weight flr.*
	In the library bookcases, tables, chairs, blinds, floors and ceilings vibrate and shake.		
	Lockers in the locker room below the mechanical room shake and rattle.		
	Rumblings and noises in library and locker room are loud.		
	The areas cannot be used by students.		
	Other parts of school effected somewhat.		
	The situation may be dangerous and cause damage to the equipment and the building.		*Close school. Shut down sys's. Resolve problem*

Assessment of Problem: Importance: urgency, size, complexity. Type problem. "State" problem is in. Risks of problem.
Extremely important, urgent, complex mechanical type problem.

Figure 6-4. Sample filled problem worksheet.

EXAMINE POSSIBLE CAUSES AND SUSPEND JUDGEMENT UNTIL VERIFIED

Cause diagnostics involve considering a thorough relevant list of possible causes of the problem (when the cause is not obvious or known), and suspending judgment until the true causes are verified.

1. ***List all the possible causes.*** Brainstorm initially and add to it as you go along. Hold back on a conclusion on the causes until the actual one(s) are validated.

2. *Initially, examine the list of possible causes and weed out obvious non-contenders.* Using judgment, logic and whatever information is available, determine which are the most likely ones, and trim the list down to the most feasible possibilities. Where possible, rate the degree of possible effect of each on the problem.

3. Gather relevant information on the possible causes. ***Make observations, examinations, tests,*** measurements and calculations, as required. Consult with others involved in the problem, and with experts as needed.

4. ***Note symptoms, syndromes, indicators and conditions.*** Note the current conditions and events. Determine the prior events and conditions, when the problem occurred. Compare to see if any have changed. Check current and prior inter-relationships of elements.

5. ***Note and describe the general and specific aspects*** involved with the causes, the *what, where, when and extent (www.e).*

6. Apply a systematic process of *elimination or confirmation* of the possible causes, by logic, evidence, obvious factors and tests. In tests, research and studies, isolate causes as required in a controlled process.

7. *Identify the relevant cause-and-effect relationships,* correlations, patterns, formulas, theories, principles, etc. Note structural, functional, spatial and time relationships.

8. Note, evaluate, and distinguish between the different characteristics of evidence—*hard factual, circumstantial, second hand, assumed, biased, prejudicial, judgmental, etc.*

9. *Consider that more than one cause may be involved. There can be: a) multiple contributing causes;* b) additive effects of causes; c) interactive effects between causes; d) new combined effect of causes.

The high school vibration problem is a perfect example of a tough cause diagnosis involving many possible causes, where it is not easily apparent if there is only one cause or many contributing causes of the vibrations and rumbling. It is also not easily distinguishable if there are interactions between the various possible causes.

CAUSE WORK SHEET

Hitting Home Runs by Wendes

Date: *September 5, 1996* Page________ of __________

PROBLEM.........(brief statement):
Vibrations in high school building.

General Description, Symptoms, etc.:
Extreme vibrations and rumbling in mechanical room being transferred to adjacent school areas.

General Area or Category	POSSIBLE CAUSES		1. Chances of being Cause, 2. Possible Degree of Effect
	General Description	Specifics	
	1. Out of balance fans.	Fans vibrating.	1:5, 33%
	2. Incorrect vibration isolators used under fans.	Some close to bottoming out.	1:5, 33%
	3. Ducts through floor connected directly to floor.	Conventional method.	1:50, 10%
	4. Locker room ceilings hung for vibrating high-pressure ducts.	Common practice.	1:50, 10%
	5. Throbbing ducts pressed against joists and beams.		Low/low
	6. Air pulsations in fan housings.		Low/low
	7. Incorrect acoustical lining used in ductwork in mechanical rm.	2 inch thick, 3 lb density.??	Low/low
	8. Debris wedged under equipment concrete inertia blocks.	Brick, wood and metal pieces.	1:1000 1%
	9. Thin lightweight concrete mechanical room floor, insufficient mass to absorb vibrations from such a large equipment load.	3 inch thick.	100:1 90%

Information, Tests, Research Needed: Priorities of testing.:

Figure 6-5. Sample filled out cause worksheet.

ESTABLISH AND EVALUATE OBJECTIVES

Develop, establish, and evaluate a thorough range of relevant objectives and conditions, which the solution must satisfy. Put in terms of what you want to achieve, maintain and avoid. This is an absolute necessity in the home run problem solving and decision-making process.

The purpose of the master list of objectives is to more accurately and fully determine your requirements, needs and wants. It is to state what the results and conditions should be of the solution, plus be able to better focus on the factors. The process is to identify, establish, describe and evaluate the requirements that must be met by the chosen solution.

The list of objectives is beneficial in a number of ways: as guide in generating better possible *solutions* and as a checklist for making the best decision. It aids in planning what to work on in the problem. It expands perspectives and viewpoints, and avoids premature conclusions.

1. *List objectives at random at first, in a brainstorming manner.* Don't be too concerned about categorizing or rating the objectives in the initial listing. Let it be more of an uncritical listing.

2. *Expand* the master list as you go along as needed. Convert problem statements, causes, pros and cons of the optional solutions, and implement problems into master list objective statements.

3. Try to make *brief phrases,* starting with an action verb and a noun, such as "increase sales" or "reduce labor time."

4. State the *overall goals or purpose* of the solution.

5. *Add specifics* to the general objective statements as needed.

6. *Identify general areas of consideration,* standard categories, criteria, features, etc. as needed.

7. *Edit* the list, if required, before comparing the optional solutions against it.

8. Evaluate the *importance* of the factors. Determine which objectives are "absolutes." These are the most critical factors relative to the success of the solution. Rate the non-absolute objectives with some scale or value.

9. Allow objectives to *digest* and for the subconscious mind to evaluate them.

OBJECTIVES WORK SHEET

Hitting Home Runs by Wendes

Date: *September, 1996*

Page ________ of ________

PROBLEM:
Vibrations in high school building

OVERALL GOAL: *Rid the building of vibrations and loud rumbling as quickly as possible, achieving at a minimum short term acceptable results, while avoiding excessive disruption of the school operations.*

General Category of Consideration	OBJECTIVES		Importance Ratings, Evaluations
	General Description	**Specifics**	
	1. *Immediate action by trades involved regardless of responsibility.*	*Inspections tests within few days.*	*Absolute requirement by school.*
	2. *Determine actual cause(s) first before full trial and error solutions.*		*Absolute by contractors.*
	3. *Know who pays for work done in the end if not their responsibility.*	*If over the normal svc amount.*	*Absolute by contractors.*
	4. *Correct without closing school down.*	*More than a few days.*	*Absolute by school.*
	5. *Correct without shutting HVAC systems down if possible, a minimum amount of time otherwise.*	*3 days in effected area.*	*Of major importance*
	6. *Avoid violating parameters of health, safety and code requirements in buildings*	*Must meet parameters 100%.*	*Absolute of school.*
	7. *Corrections should be meet an acceptable realm of the specifications, contracts, comfort, and performance requirements.*	*No greater deviation than 5 or 10%.*	*Absolute of school.*

Remarks:

Figure 6-6. Sample filled out objectives worksheet.

GENERATE MULTIPLE OPTIONS AND HOLD BACK ON FINAL CHOICE

Generating and considering multiple solutions for problems and holding back on making a choice is an absolute necessity in home run problem solving and decision-making. It greatly increases the odds of scoring the most runs, and ending up with the most beneficial solution.

1. *List optional solutions initially in a non-restrictive, brainstorming manner.*
2. Thoughts about the *effects of the various options,* about pros or cons, risks involved, and the odds of success or failure, that pop up initially or as you go along, should not be ignored. They should be simply noted, but not used to draw a final conclusion on an option. This means you shouldn't accept or reject it at that moment, unless it obviously violates an absolute objective.
3. Determine the *general plan of work,* which may be needed for generating the options. Determine if *testing or further information* is required on the development of the possible options.
4. *Expand* the list of possible options as you go along. Relate options to the problem, causes, inter-relationships, and objectives. Examine solution from similar situations. Consult with others.
5. Use your *imagination, creativity, and the subconscious mind* to generate options. Make pauses and let the subconscious mind do its job generating solutions and ideas, and clarifying things.
6. Describe options in *general and specific terms, in terms of what, where, when and extent (www.e)* as required.
7. Do an *initial rating* on the options on a rough judgmental basis and eliminate obvious poor options.
8. *Draw sketches or diagrams; make charts or graphs,* as required.
9. *Plan further* work on options, focusing and developing them on a priority basis.
10. Fill in the "*general area or category*" of the options, if this aids in understanding and organizing them better.
11. Resist *bogus options and being pressured* by others.
12. *Edit, consolidate, and reorganize the list as required,* when ready for the evaluation and decision phase.

OPTIONS WORK SHEET

Hitting Home Runs by Wendes

Date *September, 1996* Page ________ of ________

PROBLEM.........(brief statement):
Vibrations in high school.

Information and Planning Needed: *Make selective tests of possible options to determine if they are effective in reducing vibrations and noise, and determine to what degree.*

General Area or Category	OPTIONS (Possible solutions)		Ratings,
	General Description	Specifics	
	a) Balance select fans.	*Fans 1 and 6*	*Priority test.*
	b) Replace vibration isolators on fans with heavier duty type.	*All fans*	*Priority test.*
	c) Reduce total air quantities of select fans by lowering RPM's.		*Secondary test.*
	d) Isolate duct s passing through floors from the floors.	*High pressure ducts.*	*Minor*
	e) Install special sound baffles in fan housings.	*All*	*Moderate*
	f) Install air flow baffles in fan housings to get rid of pulsations.	*All*	*Moderate*
	g) Clad outside of ducts with vibration absorbing lead and sponge.	*High pressure ducts.*	*Minor*
	h) Clean out construction debris from under concrete blocks.	*All*	*Minor*
	i) Add structural supports to bldg.	*Where feasible.*	*Major*
	j) Close school and resolve problem.		*Major*

Remarks: *Tests were made on all the above items on the mechanical systems and none had any effect on reducing the vibrations. The fans were in balance, isolators correct, etc. This left some possible strong link with the light weight floor.*

Figure 6-7. Sample filled out options worksheet.

i/p

INFORMATION AND PLANNING WORK SHEET

- *Information, Research, Tests, Consulting Needed*
- *Relationships, Diagrams, Planning and Scheduling*

Hitting Home Runs, by Wendes

Date: September, 1996

Page________ of ________

PROBLEM......(brief phrase): *Vibrations in high school.*

1. The mechanical room has 30 large pieces of rotating equipment in it, fans, pumps, chillers, compressors, etc., sitting on extremely heavy concrete inertia blocks with spring isolators.
2. The floor is made up of 3-inch lightweight concrete on a metal pan.

INITIAL TESTS

3. Tests are to be made to establish the cause of the vibrations and noise by a process of elimination. A priority sequence is to be established in order of the likely hood of being the actual cause and of degree of effect.
4. Rebalance select fans.
5. Change select vibration isolators.
6. Install air flow and sound baffles in select fan housings.
7. The results of the above tests turned out to have little or no effect on the vibrations or noise levels in either the equipment room or adjacent buildings. The fans were in balance, isolators were correct, etc.

CONSULT AN EXPERT

8. Consult an expert on building vibration problems to analyze and measure the vibrating frequencies of the mechanical room floor and the equipment, and compare the two.
9. The vibration readings showed that several large fans were at close harmonic frequency with the concrete floor. This relationship of the two, in contact with each other, was producing a massive combined geometric increase in force.
10. Test turning two of the fans with a frequency closest to that of the floor off, to see what will happen.
11. Results: Ninety percent (90%) of the vibrations and rumbling stopped, thus confirming the true cause, a simultaneous harmonic frequency situation.

DECISION AND IMPLEMENTATION

12. Based on these findings, it is decided to change the drives and lower the operating speeds of the two fans about 10 to 15%.
13. Results: Ninety percent of the vibrations and rumblings in the adjacent library and locker rooms were gone. This was considered acceptable by the school and board.

Figure 6-8. The information and planning worksheet is an unrestricted form. The natural sequence of planned work and the results of the various actions can be thus recorded, as they occur. Also, use the information and planning worksheets when there is information, etc. that does not lend its self to recording on the categorized works sheets.

EVALUATE OPTIONS, COMPARE AGAINST OBJECTIVES, MAKE DECISION
(CHECKOFF LIST)

EVALUATE:

- **Effects of Optional Solutions.** Determine the possible good and bad effects of the options. Determine possible side effects and interactions.
  Some of the good effects of the harmonic frequency solution in the high school vibration situation are that it gets rid of about 90 percent of the vibrations and rumbling, it's very low cost and can be done quickly.

- **Identify and evaluate risks of the options.** Determine the risk-to-reward ratio as required.

- **Determine odds of success and of failure** of the options, and the consequences thereof.

- **Confirmations.** Confirm questionable assumptions, information, theories, opinions, reasoning, feelings, etc., which are critical to the outcome of the problem. Question principles, theories, generalities, propositions, and premises which can often be suspect. Fixed beliefs, traditions, habits, prejudices, notions and precedents can be pitfalls. Illogical reasoning, false conclusions and inferences can be risky situations. Consider the uncertainties involved with the feasible solutions.

COMPARE:

- **Compare Options to Master List of Prioritized Objectives.** Determine how well or not options meet the master list of objectives and goals. Emphasis must be on the absolute objectives, rather than on those factors of less importance.
  The harmonic frequency option meets most of the absolute requirements of the objectives 100 percent, determining true cause, immediate action, correct without closing school down, and minimal down time on two HVAC systems. It touches acceptable borderline limits on contractual, performance and code factors. Responsibility for who pays, everyone agrees, can be worked out, since the costs are minimal, and responsibility is shared among several parties.

- **Rate the finalists of the options.**

- **Make your decision accordingly.**

WRITE IT DOWN

THE POWER OF NOTES, LISTS, SKETCHES, DIAGRAMS, CHARTS

Writing down information, making notes, lists, drawings and diagrams of the factors involved in problems can be extremely beneficial in home run problem solving.

Making notes is an organized recording function for more efficient recall and effective application of the principles and procedures of the home run problem solving process.

In general notes are needed if the problem is more complex, more critical in nature or involves many factors to remember (many major sub-problems).

Benefits of making notes, lists, diagrams and sketches:

- Writing down lists of factors for the various categories of the problem solving process results in more *complete consideration of the full range of relevant factors. It aids in opening up the mind and in promoting multiplicity thinking.*
- Helps to *remember* the various factors and essential information for utilization in the evaluation and decision phases. As the Chinese proverb states, "the weakest ink lasts longer than the strongest memory."
- Allows person to *focus,* not only on the various segments of the problem solving process more easily, but also on the particular factors of the problem in each segment.
- Making notes on factors helps in *suspending judgment* until all the required preparation is completed, giving more valid *conclusions.*
- It enables better *communication* on the problem with others, and provides a *written record* for future reference, reviews, and reports.

ORGANIZED NOTES ON BLANK SHEETS OF PAPER

If notes are made on *blank sheets of paper,* they must be grouped properly, according to the segment of the problem solving process they belong to, such as information on the problem, causes, etc. Allow enough space for each category. You are not trying to conserve paper in the process, but rather make clear, complete usable notes to reach the best solution.

Association/Recall Diagrams

Notes can be made as *association/recall type diagrams*. The biggest feature of the association and recall type diagrams is that related items can be grouped easily, and items can be added to their respective groups at any time. The diagram also functions as a relatively well-organized set of notes for easier recall and communicating with others.

The diagram relates better to the *actual interconnections in the human brain*, and hence the mind works better with this visual graphical approach.

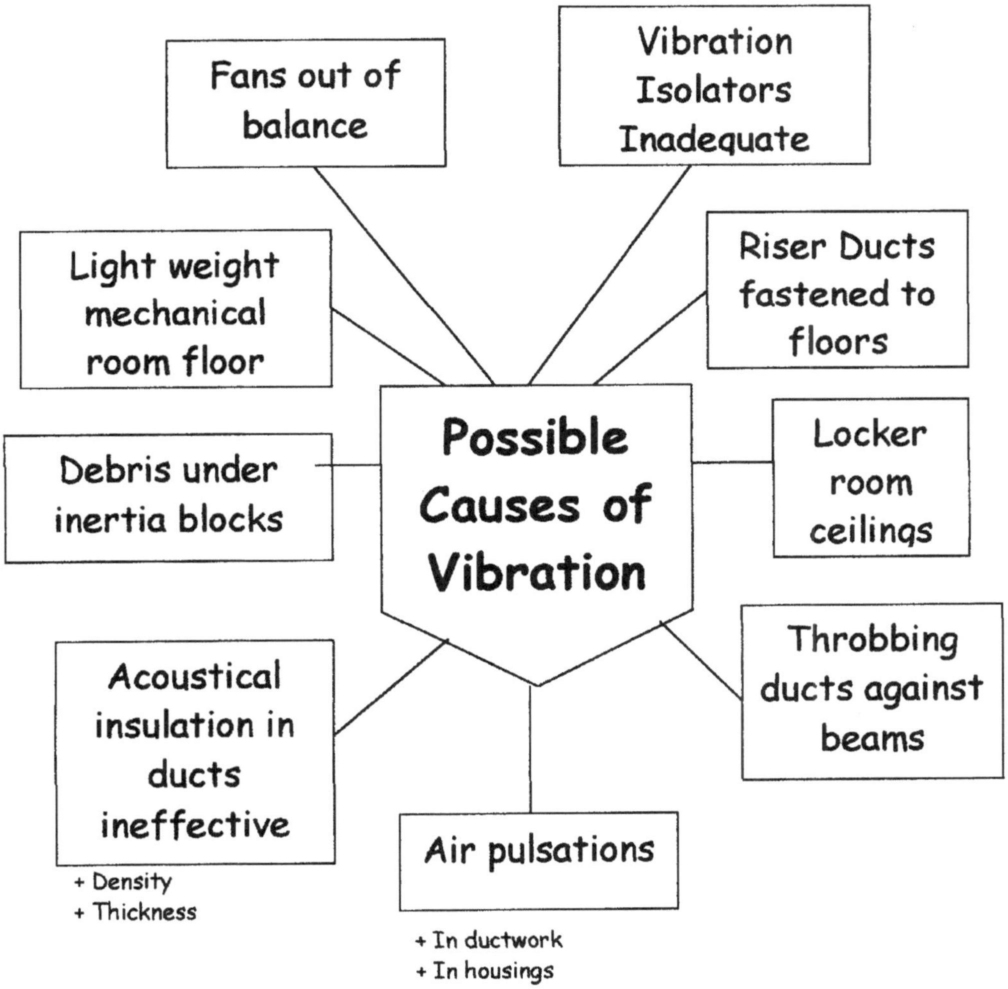

Figure 6-9. The above is an example of a association/recall type diagram for listing the possible causes of the high school vibration problem.

Bottom Line on Notes

The primary function in home run problem solving is thinking. Hence making notes must not be allowed to hinder thinking. Note taking and categorizing are only aids to thinking.

Thinking is the engine that drives the problem solving process. Making notes, however useful, is secondary.

Notes compensate for the weakness of the short-term memory, and for the inability of the human mind to handle too many factors. This limitation of the mind's subconscious over simplifies and adversely shortcuts the problem solving and decision making processes. Notes minimize chances of missing or skimming over factors, and potential confusion in the process.

Appendix A

Blank Forms

Full-size, reproducible, blank testing and balancing forms for your use.

TESTING AND BALANCING AUDIT REPORT

Date ______________________

Job ____

Location ____

Architect ________________________ Phone ________________________

Engineer _____________________ Phone ________________________

Testing and Balancing Contractor ______________________________________

Address ____

City ____________________________ State ________ Zip _____________

Phone ________________________ Fax Number __________________________

Full Size Blank Testing and Balancing Audit Forms for Your Use

GENERAL INFORMATION

Job____________________ **Job No.**__________ **Date**____________
Location______________________________

TYPE SYSTEMS

____
____
____
____
____
____
____
____

DIFFUSERS AND REGISTERS
Manufacturer______________________________
Types______________________________
____
____

TERMINAL UNITS
Manufacturer______________________________
Types______________________________
____

INSTRUMENTS USED (Indicate Models)

☐ **Velometer**__________	☐ **Flow Hood**__________
☐ **Anemometer**__________	☐ **Thermal Anemometer**__________
☐ **Manometers**__________	☐ **Magnehelics**__________
☐ **Volt-Ammeter**__________	☐ **Tachometer**__________

BUILDING AND SYSTEM COMPLETION CHECK OFF LIST

Architectural:	☐ Walls ☐ Roof ☐ Floors ☐ Windows ☐ Doors ☐ Ceilings
Electrical:	☐ Starters ☐ Overload ☐ Transformers Tested ☐ Wiring
Controls:	☐ Control Motors ☐ Linkages ☐ Compressors ☐ Stats ☐ Tubing ☐ Wiring
Piping:	☐ Coils ☐ Valves ☐ Piping ☐ Pumps ☐ Wiring
Sheet Metal:	☐ Grilles ☐ Fins ☐ Drives ☐ Air Handling Units ☐ Filters ☐ Wiring

Remarks______________________________
____

Full Size Blank Testing and Balancing Audit Forms for Your Use (Continued)

FAN TEST REPORT

Job ______________________ Job No ______ Date ______________

Location ______________________ System ______________

Equipment Location ______________ Serves ______________ Tested By: ______________

☐ Air Handling Unit ☐ Roof Top Unit ☐ Furnace ☐ Supply Fan ☐ Exhaust Fan ☐ Pkg Unit

☐ LP ☐ MP ☐ HP ☐ Constant Volume ☐ VAV

FAN DATA	
Manufacturer	
Model Size	
Type Fan	☐ Centrigal ☐ Roof Exhaust ☐ Inline ☐ Vane Axial ☐ Prop.
Type Wheel	☐ Backward Incline ☐ Air Foil ☐ Forward Curve ☐ Paddle Wheel
Wheel: ☐ Alignment OK ☐ Gap ☐ Fastened ☐ Clean	
Belts	C to C Distance
Pulleys: Fan Dia.	Mot. Dia.
Motor Movement	
Bearings ☐ Zerk ☐ Seal ☐ Cut Off Plate OK	

MOTOR		
Manufacturer	Serial No.	
Frame No.	Type	☐ T ☐ U
Svc. Fact.	Rated	Actual
HP, Nameplate		
BHP		
Amps, L_1 L_2 L_3		
Voltage, L_1 L_2 L_3		
RPM		
Phase		

FAN PERFORMANCE	Design	Actual
Fan CFM		
Outlet CFM Total		
Fan RRM		
Fan S.P.		

STARTER	
Manufacturer	Model
Starter Size	Class
Overload: Required Size	
Actual	

CONDITIONS			
Vortex Damper Position			
Outside Air Damper Setting			
Return Air Damper Setting			
Filter Conditions			
Coil Conditions			
Temperatures			
OA:	DB	WB	RH
RA:	DB	WB	RH
Mixed Air:	DB	WB	RH
Discharge	DB	WB	RH
Space:	DB	WB	RH
Duct Temp. Drop	DB		

STATIC PRESSURE DROPS	Upstream	Downstream	Total Drop
Filter			
Heat. Coil			
Cool. Coil			
Fan Inlet			
Fan Discharge			
Total Fan S.P.			

Remarks

Wendes Engineering and Contracting Services 7/80 Form TAB 201

Full Size Blank Testing and Balancing Audit Forms for Your Use (Continued)

OUTLET AIR BALANCE REPORT

Project ____________________ Job No. __________ Date ____________

Location ______________________________ System ____________

Instruments Used ______________________ Tested by: ____________

ROOM AREA SERVED	OPENING				REQUIRED		PRELIMINARY			FINAL	
	No.	Model	Size	A_k	Vel	CFM				Vel	CFM

Remarks ____

____

____

Wendes Engineering and Contracting Services Form No. TAB202

Full Size Blank Testing and Balancing Audit Forms for Your Use (Continued)

HVAC SYSTEMS RECAP SHEET

Job ________________ Job No. ________ Date ________________

Location ________________ ☐ Air ☐ Water Balancer ________________

SYSTEM	LOCATION OF EQUIPMENT	STATUS			CFM or GPM			RPM		S.P. or Ft Head		AMPS		TONS	MBH	REMARKS
		1	2	3	Design	Actual	Percent of Design	Design	Act	Des	Act	Des	Act	Cool	Heat	

1. System ready for balancing 2. Equipment checked out 3. System balanced

Wendes Engineering and Contracting Services

7/80 Form TAB 206

Full Size Blank Testing and Balancing Audit Forms for Your Use (Continued)

PITOT TUBE TRAVERSE, RECTANGULAR DUCT

Job ______________________________ Job No. ________ Date ____________

Location ______________________________ System ____________

Location of Duct ______________________________ ℄ SP ____________

Duct Size		Required CFM	Required FPM
Duct Area	Sq Ft	Actual CFM	Actual FPM

Percent of Design: = Actual CFM / Required CFM = ( ) / ( ) = ________

☐ Initial
☐ Final
Readings

	1	2	3	4	5	6	7	8	
	VP / FPM	VP / FPM	VP / FPM	VP / FPM	VP / FPM	VP / FPM	VP / FPM	VP / FPM	
A	○	○	○	○	○	○	○	○	Y/2
B	○	○	○	○	○	○	○	○	Y
C	○	○	○	○	○	○	○	○	
D	○	○	○	○	○	○	○	○	
E	○	○	○	○	○	○	○	○	
F	○	○	○	○	○	○	○	○	
Column Totals →									= Total of all Columns

X/2 X

DUCT WIDTH

1. Divide recommended spacing into duct width for **equal spacing** between points of readings and write in each space.
2. Split remainder in half for **spacing at sides** of duct and write in.
3. To determine the **actual average FPM**, total the FPM in each column, add the columns together and divide by total number of points in traverse.

(TOTAL ALL COLUMNS) / (Total No. of points) = ( ) / ( ) = [ ] Average FPM

RECOMMENDED EQUAL SPACING BETWEEN POINTS OF READING	
Width Range of Duct Inches	Spacing in inches
8-14	5
15-24	6
25-36	7
37-48	8
49-60	9
61-72	10
73-64	11
85-96	12

Wendes Engineering and Contracting Services

CONVERTING VP INTO FPM

VP	FPM	VP	FPM	VP	FPM	VP	FPM	VP	FPM
.01	400	.31	2230	.61	3127	.91	3821	1.21	4405
.02	566	.32	2260	.62	3153	.92	3842	1.22	4423
.03	693	.33	2301	.63	3179	.93	3863	1.23	4442
.04	801	.34	2335	.64	3204	.94	3884	1.24	4460
.05	895	.35	2369	.65	3229	.95	3904	1.25	4478
.06	981	.36	2403	.66	3254	96	3924	1.26	4495
.07	1060	.37	2436	.67	3279	.97	3945	1.27	4513
.08	1133	.38	2469	68	3303	.98	3965	1.28	4531
.09	1201	.39	2501	.69	3327	.99	3985	1.29	4549
.10	1266	.40	2533	.70	3351	1.00	4005	1.30	4566
.11	1328	.41	2563	.71	3375	1.01	4025	1.31	4583
.12	1387	.42	2595	.72	3398	1.02	4045	1.32	4601
.13	1444	.43	2626	.73	3422	1.03	4064	1.33	4619
.14	1498	.44	2656	.74	3445	1.04	4084	1.34	4636
.15	1551	.45	2687	.75	3468	1.05	4103	1.35	4653
.16	1602	.46	2716	.76	3491	1.06	4123	1.36	4671
.17	1651	.47	2746	.77	3514	1.07	4142	1.37	4688
.18	1699	.48	2775	.78	3537	1.08	4162	1.38	4705
.19	1746	.49	2804	.79	3560	1.09	4181	1.39	4722
.20	1791	.50	2832	.80	3582	1.10	4200	1.40	4739
.21	1835	.51	2860	.81	3604	1.11	4219	1.41	4756
.22	1879	.52	2888	.82	3625	1.12	4238	1.42	4773
.23	1921	.53	2916	.83	3657	1.13	4257	1.43	4790
.24	1962	.54	2943	.84	3669	1.14	4276	1.44	4806
.25	2003	55	2970	.85	3690	1.15	4295	1.45	4823
.26	2042	.56	2997	.86	3709	1.16	4314	1.46	4840
.27	2081	.57	3024	.87	3729	1.17	4332	1.47	4856
.28	2119	.58	3050	.88	3758	1.18	4350	1.48	4873
.29	2157	.59	3076	.89	3779	1.19	4368	1.49	4889
.30	2193	.60	3102	.90	3800	1.20	4386	1.50	4905

7/80 Form TAB 203

Full Size Blank Testing and Balancing Audit Forms for Your Use (Continued)

PITOT TUBE TRAVERSE, SMALL ROUND DUCTS

Job ______________________ Job No. ________ Date ____________

Location ______________________________________ System ____________

Location of Duct ________ Area Served ________ Duct Temperature ________ ℄ SP ________

Duct Diameter	Required CFM	
Duct Area Sq Ft	Actual CFM	Actual FPM

Percent of Design: = Actual CFM / Required CFM = ( ) / ( ) = ____________ percent

TRAVERSE POINT LAYOUT

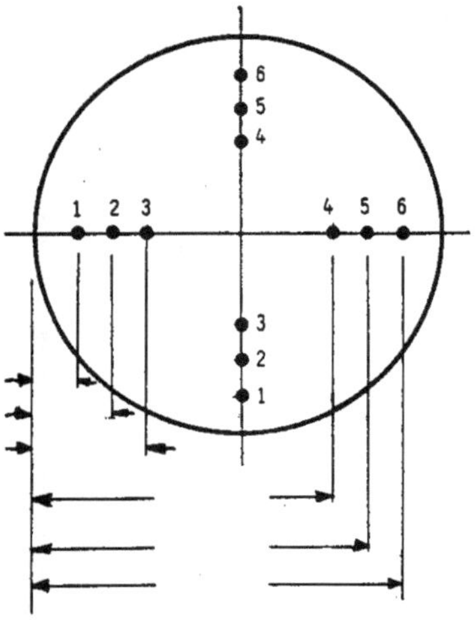

TRAVERSE POINTS	READINGS			
	At Startup		Final	
	V.P.	FPM	V.P.	FPM
1				
2				
3				
4				
5				
6				
1				
2				
3				
4				
5				
6				
Total FPM				
Divided by 12 =				
Times Duct Area				
= Total CFM				

Distance from side of duct to point of reading to nearest 1/8th inch for 6 point traverse.

	POINTS					
Duct Dia.	1	2	3	4	5	6
3	1/8	1/2	7/8	2 1/8	2 1/2	2 7/8
4	1/8	5/8	1 1/8	2 7/8	3 3/8	3 7/8
5	1/4	3/4	1 1/2	3 1/2	4 1/4	4 3/4
6	1/4	7/8	1 3/4	4 1/4	5 1/8	5 3/4
7	1/4	1"	2"	5"	6"	6 5/8
8	3/8	1 1/8	2 3/8	5 5/8	6 3/4	7 5/8
9	3/8	1 3/8	2 5/8	6 3/8	7 5/8	8 5/8
10	3/8	1 1/2	3"	7"	8 1/2	9 1/2

CONVERTING VP INTO FPM

VP	FPM	VP	FPM	VP	FPM	VP	FPM	VP	FPM
.01	400	.31	2230	.61	3127	.91	3821	1.21	4405
.02	566	.32	2260	.62	3153	.92	3842	1.22	4423
.03	693	.33	2301	.63	3179	.93	3863	1.23	4442
.04	801	.34	2335	.64	3204	.94	3884	1.24	4460
.05	895	.35	2369	.65	3229	.95	3904	1.25	4478
.06	981	.36	2403	.66	3254	.96	3924	1.26	4495
.07	1060	.37	2436	.67	3279	.97	3945	1.27	4513
.08	1133	.38	2469	.68	3303	.98	3965	1.28	4531
.09	1201	.39	[illegible]	.69	3327	.99	3985	1.29	4549
.10	1266	.40	2533	.70	3351	1.00	4005	1.30	4566
.11	1328	.41	2563	.71	3375	1.01	4025	1.31	4583
.12	1387	.42	2595	.72	3398	1.02	4045	1.32	4601
.13	1444	.43	2626	.73	3422	1.03	4064	1.33	4619
.14	1498	.44	2656	.74	3445	1.04	4084	1.34	4636
.15	1551	.45	2687	.75	3468	1.05	4103	1.35	4653
.16	1602	.46	2716	.76	3491	1.06	4123	1.36	4671
.17	1651	.47	2746	.77	3514	1.07	4142	1.37	4688
.18	1699	.48	2775	.78	3537	1.08	4162	1.38	4705
.19	1746	.49	2804	.79	3560	1.09	4181	1.39	4722
.20	1791	.50	2832	.80	3582	1.10	4200	1.40	4739
.21	1835	.51	2860	.81	3604	1.11	4219	1.41	4756
.22	1879	.52	2888	.82	3625	1.12	4238	1.42	4773
.23	1921	.53	2916	.83	3657	1.13	4257	1.43	4790
.24	1962	.54	2943	.84	3669	1.14	4276	1.44	4806
.25	2003	.55	2970	.85	3690	1.15	4295	1.45	4823
.26	2042	.56	2997	.86	3709	1.16	4314	1.46	4840
.27	2081	.57	3024	.87	3729	1.17	4332	1.47	4856
.28	2119	.58	3050	.88	3758	1.18	4350	1.48	4873
.29	2157	.59	3076	.89	3779	1.19	4368	1.49	4889
.30	2193	.60	3102	.90	3800	1.20	4386	1.50	4905

Wendes Engineering and Contracting Services

7/80 Form TAB 204a

Full Size Blank Testing and Balancing Audit Forms for Your Use (Continued)

PITOT TUBE TRAVERSE, LARGE ROUND DUCTS

Job__________ Job No.__________ Date__________

Location__________ System__________

Location of Duct__________ Area Served__________ Duct Temperature__________ ℄ SP__________

Duct Diameter	Required CFM	Actual CFM
Duct Area Sq Ft		Actual FPM

Percent of Design: = Actual CFM / Required CFM = ( ) / ( ) = __________ percent

TRAVERSE POINT LAYOUT

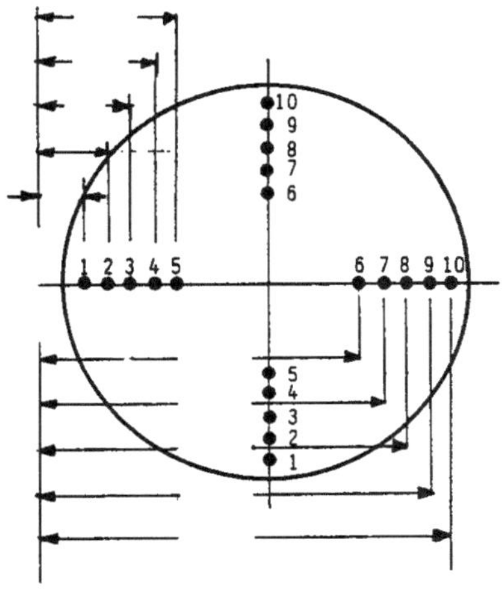

Distance from side of duct to point of reading to nearest 1/8th inch for 10 point traverse.

DUCT DIA.	1	2	3	4	5	6	7	8	9	10
12	3/8	1	1-3/4	2-3/4	4-1/8	7-7/8	9-1/4	10-1/4	11	11-5/8
13	3/8	1	1-7/8	2-7/8	4-1/2	8-1/2	10-1/8	11-1/8	12	12-5/8
14	3/8	1-1/8	2	3-1/8	4-3/4	9-1/4	10-7/8	12	12-7/8	13-5/8
15	3/8	1-1/4	2-1/4	3-3/8	5-1/8	9-7/8	11-5/8	12-3/4	13-3/4	14-5/8
16	3/8	1-1/4	2-3/8	3-5/8	5-1/2	10-1/2	12-3/8	13-5/8	14-3/4	15-5/8
17	1/2	1-3/8	2-1/2	3-7/8	5-3/4	11-1/4	13-1/8	14-1/2	15-5/8	16-1/2
18	1/2	1-1/2	2-5/8	4-1/8	6-1/8	11-7/8	13-7/8	15-3/8	16-1/2	17-1/2
19	1/2	1-1/2	2-3/4	4-1/4	6-1/2	12-1/2	14-3/4	16-1/4	17-1/2	18-1/2
20	1/2	1-5/8	2-7/8	4-1/2	6-7/8	13-1/8	15-1/2	17-1/8	18-3/8	19-1/2
22	5/8	1-3/4	3-1/4	5	7-1/2	14-1/2	17	18-3/4	20-1/4	21-3/8
24	5/8	2	3-1/2	5-1/2	8-1/4	15-3/4	18-1/2	20-1/2	22	23-3/8
26	5/8	2-1/8	3-3/4	5-7/8	8-7/8	17-1/8	20-1/8	22-1/4	23-7/8	25-3/8
28	3/4	2-1/4	4-1/8	6-3/8	9-5/8	18-3/8	21-5/8	23-7/8	25-3/4	27-1/4
30	3/4	2-1/2	4-3/8	6-3/4	10-1/4	19-3/4	23-1/4	25-5/8	27-1/2	29-1/4
32	7/8	2-5/8	4-5/8	7-1/4	11	21	24-3/4	27-3/8	29-3/8	31-1/8
34	7/8	2-3/4	5	7-3/4	11-5/8	22-3/8	26-1/4	29	31-1/4	33-1/8
36	1	3	5-1/4	8-1/8	12-3/8	23-5/8	27-7/8	30-3/4	33	35
38	1	3-1/8	5-1/2	8-5/8	13	25	29-3/8	32-1/2	34-7/8	37
40	1	3-1/4	5-7/8	9	13-5/8	26-3/8	31	34-1/8	36-3/4	39
42	1-1/8	3-3/8	6-1/8	9-1/2	14-3/8	27-5/8	32-1/2	35-7/8	38-5/8	40-1/8
44	1-1/8	3-5/8	6-3/8	10	15	29	34	37-5/8	40-3/8	42-7/8
46	1-1/4	3-3/4	6-3/4	10-3/8	15-3/4	30-1/4	35-5/8	39-1/4	42-1/4	44-3/4
48	1-1/4	4	7	10-7/8	16-3/8	31-5/8	37-1/8	41	44	46-3/4

TRAVERSE POINTS	READINGS			
	At Startup		Final	
	V.P.	FPM	V.P.	FPM
1				
2				
3				
4				
5				
6				
7				
8				
9				
10				
1				
2				
3				
4				
5				
6				
7				
8				
9				
10				

Total FPM		X	
Divided by 20 =			
Times Duct Area			
= Total CFM			

Wendes Engineering and Contracting Services

Form TAB 204b

Full Size Blank Testing and Balancing Audit Forms for Your Use (Continued)

PUMP TEST REPORT

Job ______________________ Job No. ________ Date ________

Location ______________________ System ________

Equipment Location ________ Serves ________ Tested by: ________

PUMP DATA		
Manufacturer		
Model /Size		
Type Pump		
Impeller Size		
	Rated	Actual
GPM		
Total Ft.Head		
RPM		

MOTOR		
Manufacturer Serial No.		
Frame No. Svc. Factor		
	Rated	Actual
HP, Nameplate		
Amps, L_1 L_2 L_3		
Voltage, L_1 L_2 L_3		
RPM		
Phase		

PUMP PRESSURES		
	Design	Actual
Static Hd(Pump Off)		
Discharge		
Suction		
Block Off: (Running, no flow)		
Discharge		
Suction		
Total		
Running:		
Discharge		
Suction		
Total		

STARTER	
Manufacturer	Model
Size	Class
Overload: Required Size :	
Actual:	

BHP $[HP_{np} \times \frac{Aa}{Ar} \times \frac{Va}{Vr}]$

$$\text{KWH Per Year} = \frac{\text{Volts} \times \sqrt{3}^{*} \times \text{Avg Amps} \times \text{Yearly Hours of Operation} \times \text{PF}}{1000}$$

= [ ] *(3 phase)

Remarks ______________________

Wendes Engineering and Contracting Services 7/80 Form TAB 210

Full Size Blank Testing and Balancing Audit Forms for Your Use (Continued)

WATER BALANCE REPORT

Flow or Pressure Drop

Job ______________________ Job No. ________ Date ____________

Location ______________________ System ____________

Instrument Used ______________________ Tested by: ____________

TAG NO. ITEM	SIZE	REQUIRED		BY PASS		PRELIMINARY						FINAL	
		ΔP[1]	GPM	ΔP	GPM	ΔP	%	ΔP	%	ΔP	%	ΔP	GPM

1. ΔP: Can represent flow station differential or pressure drop across item.

Remarks ______________________

Wendes Engineering and Contracting Services 7/80 Form TAB 211

Full Size Blank Testing and Balancing Audit Forms for Your Use (Continued)

WATER BALANCE REPORT

Temperature Measurements

Job ____________ Job No. ______ Date ______

Location ____________ System ______

Instrument Used ____________ Tested by: ______

COIL	SIZE	DESIGN				TEST				FINAL			
		EWT	LWT	EAT	LAT	EWT	LWT	EAT	LAT	EWT	LWT	EAT	LAT

Remarks ____________

Wendes Engineering and Contracting Services 7/80 Form TAB 212

Full Size Blank Testing and Balancing Audit Forms for Your Use (Continued)

CHILLER TEST REPORT

Job ______________ Job No. ________ Date ________

Location ______________ System ________

Equipment Location ________ Serves ________ Tested by: ________

COMPRESSOR DATA
Manufacturer
Model/Size
Type
Capacity tons @ GPM
Refrigerant Pounds
KW KW Per Ton
Serial No.

COMPRESSOR MOTOR		
Manufacturer Serial No.		
Frame No. Type Frame ☐T ☐U		
Svc. Factor:	Rated	Actual
HP, Nameplate		
BHP $[HPnp \times \frac{Aa}{Ar} \times \frac{Va}{Vr}]$		
Amps, L_1 L_2 L_3		
Voltage, L_1 L_2 L_3		
RPM		
Phase		

COMPRESSOR	Design	Actual
Suction Pressure		
Suction Temp.		
Discharge Press.		
Discharge Temp.		
Oil Temp/Press.		

STARTER
Manufacturer Model
Size Class
Overload: Required Size :
Actual:

EVAPORATOR	Design	Actual
Refrig. Pressure		
Refrig. Temp.		
Ent. Water Pressure		
Lvg. Water Pressure		
Ent. Water Temp.		
Lvg. Water Temp.		
Flow GPM		

CONDENSER	Design	Actual
Liquid Line Pressure		
Liquid Line Temp.		
Ent. Water Press.		
Lvg. Water Press.		
Ent. Water Temp.		
Lvg. Water Temp.		
Flow GPM		

CONDITIONS
Refrigerant Level
Oil Level
Percent Cylinders Unloaded
Chilled Wat. Control Setting
Condenser Wat. Control Setting
Low Wat. Cutout Temp. Setting
Low Pressure Cutout Setting
High Pressure Cutout Setting

$$\text{KWH Per Year} = \frac{\text{Volts} \times 1.73 \times \text{Avg Amps} \times \text{Yearly Hours of Operation} \times \text{PF}}{1000}$$

KW's = ________
Per Year

Remarks ________

☐ Purge Operation Checked ________

☐ Crankcase Heater Checked ________

____

____

____

Wendes Engineering and Contracting Services

7/80 Form TAB 214

Full Size Blank Testing and Balancing Audit Forms for Your Use (Continued)

HVAC ENERGY AUDITING REPORT

Date ______________________

Job ____

Location ____

Architect ______________________ Phone ______________________

Engineer ______________________ Phone ______________________

Auditing Contractor ____

Address ____

City ______________________ State __________ Zip ______________

Phone ______________________ Fax ______________________

BUILDING AND SYSTEM DESCRIPTION

Name ______________________________

Location ______________________________

Latitude __________ Elevation __________ When Built __________

A. CATEGORY OF STRUCTURE

____

B. BUILDING DESCRIPTION

Area, Sq Ft: __________ Number of Floors: __________

Volume, Cu Ft: ______________________________

Number of Occupants: __________ Sq Ft/Person: __________

Types of Areas: ______________________________

____

____

C. CONSTRUCTION DETAILS

Glass: ______________________________

____

Exterior Walls: ______________________________

____

Roof and Ceilings: ______________________________

____

Floors: ______________________________

Total Exposed Wall Area Sq Ft: ______________________________

Total Glass Area Sq Ft: __________ Percent __________

D. HOURS OF OCCUPANCY AND OPERATION

Working Hours: ______________________________

Lighting Hours: ______________________________

HVAC Hours: ______________________________

Janitorial Cleanup Times: ______________________________

Computer Room: ______________________________

Other: ______________________________

E. HEATING AND COOLING SYSTEMS DESCRIPTION

____

____

____

____

____

____

Wendes Mechanical Consulting Services Co. Form AUD 301

Energy Audit Forms for Your Use (Continued)

F. ANNUAL ENERGY CONSUMPTION

Total Heating, Cooling, Electrical, Lighting Per Yr:

Total BTU: ____________ BTU Per Sq Ft: ____________

Total Energy Costs: ____________ Costs Per Sq Ft: ____________

Electrical, Total KWH: ____________ KWH/Sq Ft: ____________

Total Elec. Costs: ____________ Costs/Sq Ft: ____________

Heating Fuels, BTU Per Yr: ____________ Per Sq Ft: ____________

Total Fuel Costs: ____________ Costs/Sq Ft: ____________

G. ORIGINAL ENVIROMENTAL DESIGN CONDITIONS

Heating

Peak Heat Loss BTUH: ____________ Degree Days: ____________

Design Temperatures: ____________

Avg Winter Temp.: ____________ Avg Winter Hours: ____________

Cooling

Peak Heat Gain BTUH: ____________ Degree Days: ____________

Design Temperatures: ____________

____

Avg Summer Temp.: ____________ Avg Summer Cool Hours: ____________

Air and Hydronic Flows

Supply CFM: ____________ CFM/Sq Ft: ____________

Exhaust Air CFM: ____________ Exh Air/Sq Ft: ____________

Min Outside Air CFM: ____________ OA Per Person, Sq Ft: ____________

Make Up Air CFM: ____________

HVAC GPM: ____________ Domestic GPM: ____________

H. LIGHTING

Levels in Foot Candles: ____________

Levels in Watts/Sq Ft: ____________

Type: ____________

I. ELECTRICAL SERVICE

Type: ____________ Metering: ____________

Voltage: ____________

J. CONNECTED ELECTRICAL LOADS (KW's)

Lighting: ____________ Office Equipment: ____________

Heating and Cooling Equipment: ____________

Air Handling and Exhausts: ____________

Cooking: ____________ Machinery: ____________

Wendes Mechanical Consulting Services Co.

Full Size Blank Testing and Balancing Audit Forms for Your Use (Continued)

ELECTRICAL CONSUMPTION HISTORY

BUILDING ______________________ YEAR ______________________

______________________ SIZE SQ FT ______________________

MONTH	NO OF DAYS	KWH USED	COST PER KWH	ELECTRICAL COSTS DEMAND PEAK	ELECTRICAL COSTS DEMAND CHARGE	POWER FACTOR ADJ	FUEL ADJ	TOTAL COST
JAN								
FEB								
MAR								
APR								
MAY								
JUNE								
JULY								
AUG								
SEPT								
OCT								
NOV								
DEC								
TOTAL								
AVG/MO								

Wendes Mechanical Consulting Services Co. Form AUD 301

Full Size Blank Testing and Balancing Audit Forms for Your Use (Continued)

FAN TEST REPORT

Job ______ Job No ______ Date ______

Location ______ System ______

Equipment Location ______ Serves ______ Tested By: ______

☐Air Handling Unit ☐Roof Top Unit ☐Furnace ☐Supply Fan ☐Exhaust Fan ☐Pkg Unit

☐LP ☐MP ☐HP ☐Constant Volume ☐VAV

FAN DATA	
Manufacturer	
Model Size	
Type Fan	☐Centrigal ☐Roof Exhaust ☐Inline ☐Vane Axial ☐Prop.
Type Wheel	☐Backward Incline ☐Air Foil ☐Forward Curve ☐Paddle Wheel
Wheel: ☐Alignment OK ☐Gap ☐Fastened ☐Clean	
Belts	C to C Distance
Pulleys: Fan Dia.	Mot. Dia.
Motor Movement	
Bearings ☐Zerk ☐Seal ☐Cut Off Plate OK	

MOTOR		
Manufacturer	Serial No.	
Frame No.	Type	☐T ☐U
Svc. Fact.	Rated	Actual
HP, Nameplate		
BHP		
Amps, L_1 L_2 L_3		
Voltage, L_1 L_2 L_3		
RPM		
Phase		

FAN PERFORMANCE	Design	Actual
Fan CFM		
Outlet CFM Total		
Fan RRM		
Fan S.P.		

STARTER	
Manufacturer	Model
Starter Size	Class
Overload: Required Size	
Actual	

STATIC PRESSURE DROPS	Upstream	Downstream	Total Drop
Filter			
Heat. Coil			
Cool. Coil			
Fan Inlet			
Fan Discharge			
Total Fan S.P.			

CONDITIONS			
Vortex Damper Position			
Outside Air Damper Setting			
Return Air Damper Setting			
Filter Conditions			
Coil Conditions			
Temperatures			
OA:	DB	WB	RH
RA:	DB	WB	RH
Mixed Air:	DB	WB	RH
Discharge	DB	WB	RH
Space:	DB	WB	RH
Duct Temp. Drop	DB		

Remarks

Wendes Engineering and Contracting Services 7/80 Form TAB 201

Full Size Blank Testing and Balancing Audit Forms for Your Use (Continued)

PUMP TEST REPORT

Job ______ Job No. ______ Date ______

Location ______ System ______

Equipment Location ______ Serves ______ Tested by: ______

PUMP DATA		
Manufacturer		
Model/Size		
Type Pump		
Impeller Size		
	Rated	Actual
GPM		
Total Ft.Head		
RPM		

MOTOR		
Manufacturer Serial No.		
Frame No. Svc. Factor		
	Rated	Actual
HP, Nameplate		
Amps, L_1 L_2 L_3		
Voltage, L_1 L_2 L_3		
RPM		
Phase		

PUMP PRESSURES		
	Design	Actual
Static Hd(Pump Off)		
Discharge		
Suction		
Block Off: (Running, no flow)		
Discharge		
Suction		
Total		
Running:		
Discharge		
Suction		
Total		

STARTER
Manufacturer Model
Size Class
Overload: Required Size :
Actual:

BHP $[HP_{np} \times \frac{A_a}{A_r} \times \frac{V_a}{V_r}]$

$$\text{KWH Per Year} = \frac{\text{Volts} \times \sqrt{3}^{*} \times \text{Avg Amps} \times \text{Yearly Hours of Operation} \times PF}{1000}$$

= ______ *(3 phase)

Remarks ______

Wendes Engineering and Contracting Services

7/80 Form TAB 210

Full Size Blank Testing and Balancing Audit Forms for Your Use (Continued)

CHILLER TEST REPORT

Job ______ Job No. ______ Date ______

Location ______ System ______

Equipment Location ______ Serves ______ Tested by: ______

COMPRESSOR DATA		
Manufacturer		
Model/Size		
Type		
Capacity	tons @	GPM
Refrigerant	Pounds	
KW	KW Per Ton	
Serial No.		

COMPRESSOR MOTOR		
Manufacturer	Serial No.	
Frame No.	Type Frame ☐T ☐U	
Svc. Factor:	Rated	Actual
HP, Nameplate		
BHP $[HPnp \times \frac{Aa}{Ar} \times \frac{Va}{Vr}]$		
Amps, L_1 L_2 L_3		
Voltage, L_1 L_2 L_3		
RPM		
Phase		

COMPRESSOR	Design	Actual
Suction Pressure		
Suction Temp.		
Discharge Press.		
Discharge Temp.		
Oil Temp/Press.		

STARTER	
Manufacturer	Model
Size	Class
Overload: Required Size :	
Actual:	

EVAPORATOR	Design	Actual
Refrig. Pressure		
Refrig. Temp.		
Ent. Water Pressure		
Lvg. Water Pressure		
Ent. Water Temp.		
Lvg. Water Temp.		
Flow GPM		

CONDENSER	Design	Actual
Liquid Line Pressure		
Liquid Line Temp.		
Ent. Water Press.		
Lvg. Water Press.		
Ent. Water Temp.		
Lvg. Water Temp.		
Flow GPM		

CONDITIONS
Refrigerant Level
Oil Level
Percent Cylinders Unloaded
Chilled Wat. Control Setting
Condenser Wat. Control Setting
Low Wat. Cutout Temp. Setting
Low Pressure Cutout Setting
High Pressure Cutout Setting

$$\text{KWH Per Year} = \frac{\text{Volts} \times 1.73 \times \text{Avg Amps} \times \text{Yearly Hours of Operation} \times \text{PF}}{1000}$$

KW's = ______
Per Year

Remarks ______

☐ Purge Operation Checked

☐ Crankcase Heater Checked

______

______

______

Wendes Engineering and Contracting Services 7/80 Form TAB 214

Full Size Blank Testing and Balancing Audit Forms for Your Use (Continued)

ELECTRICAL LOADS

☐ EXISTING ☐ NEW

Building ____ __________Date__________________________

____

ITEM	HP OR KWH	RATED VOLTS	RATED AMPS	PH	HOURS OF OPERATION PER YEAR	ACTUAL AVERAGE AMP LOAD	ACTUAL AVERAGE VOLT LOAD	KWH PER YEAR	COSTS PER YEAR

$$KWH/YEAR = \frac{1.73 \times I \times E \times HOURS}{1000}$$

COSTS PER KWH = $0.07

ides Mechanical Consulting Services Co.

Form AUD 305

Full Size Blank Testing and Balancing Audit Forms for Your Use (Continued)

ELECTRICAL LOADS RECAP

[X] EXISTING   [ ] NEW

BUILDING SUBURBAN OFFICE BUILDING   Date 11-18-1985

AIR HANDLING EQUIPMENT   X Year Around   Summer   Winter   X Peak

ITEM	CONNECTED LOAD KWH	CONNECTED LOAD HP	PH	HOURS OF OPERATION PER YEAR	ACTUAL LOAD KWH	ACTUAL LOAD BHP	ACTUAL KWH PER YEAR	COSTS PER YEAR	PERCENT OF TOTAL
TOTAL									

1 HP = 746 WATTS   COSTS PER KWH = $0.070

Wendes Mechanical Consulting Services Co.   Form AUD 306

Full Size Blank Testing and Balancing Audit Forms for Your Use (Continued)

HEAT LOSS CALCULATION

PEAK PER HR ---- SEASONAL ---- EXISTING ---- NEW ----

Building ______________________ Date ______________

Location ______________________ Latitude ______________

Type Building ______________ Stories ________ When Built ________

Sq Ft Area ______________________ Cubic Ft of Space ______________

Calculation for Whole Building ----- For Partial Area ----- ______________

Budget Load: Sq Ft ____________ x BTUYR/Sq Ft ________ = ____________

Outside Design: DB ________ WB ________ Avg· OA Temp. ________ Winter Hours ________

Inside Design: DB(day) ________ RH ________ DB(night) ________

ITEM	DIMENSIONS	SQ FT	U	TEMP DIFF	BTU: Per Hour Seasonal	SEASONAL HOURS
ROOF OR CEILING						
FLOOR						
GLASS						
DOORS						
WALLS						
COLD INSIDE WALLS						
VENTILATION						
DUCT LOSSES						
GROSS TOTAL HEAT LOSS						
HEAT GAINS; LITES						
PEOPLE	No. People =					
OFF. EQUIPMENT	HP =					
COMPUTERS	KW =					
TOTAL INTERNAL GAINS						
NET TOTAL BUILDING HEAT LOSS						
INPUT TO HEATING EQUIPMENT, efficiency =						

BTUH = Sq Ft x U x Temp. Diff. BTU/YEAR = Sq Ft x U x Avg Temp. Diff. x Winter Hours

* BTU/SQ FT/YEAR

Remarks ______________________________

Wendes Mechanical Consulting Services Co. Form ENG 401

Full Size Blank Testing and Balancing Audit Forms for Your Use (Continued)

COOLING LOAD CALCULATION

PEAK PER HR ____ SEASONAL ____ EXISTING ____ NEW ____

Building ______________________ Date ______________

Location ______________________ Latitude _______ Peak Load, HR, MO: ____________

Type Building ______________________ Stories ____________ When Built ______________

Sq Ft Area ______________________ Cubic Ft of Space ______________

X Calculation for Whole Building ______ For Partial Area ______________

Building Load: Sq Ft ______________ x Sq Ft/Ton ______________ = ______________

Outside Design: DB _________ WB _______ Avg OA Temp. ________ Summer Hours ______________

Inside Design: DB(day) ______________ RH _________ DB(night) ____________

ITEM (Orient.)	DIMENSIONS	SQ FT	U or FACTOR*	TEMP DIFF	SENSIBLE BTUH	TONS	LATENT BTUH
ROOF OR CEILING							
GLASS							
(Solar)							
GLASS							
(Conduction)							
WALLS							
(Conduction)							
LIGHTING							
PEOPLE, NO.	(NO.)						
OFF. EQUIPMENT	MOTORS (HP)						
COMPUTERS	(KW)						
KITCHEN							
VENTIL. AIR, Sens.							
VENTIL. AIR, Lat.							
TOTAL COOLING LOAD,			SENSIBLE				
			LATENT				
			GRAND TOTAL				

BTUH = Sq Ft x U x Temp. Diff.
* U, CLTD, CLF, KW, HP, WD
CLTD = (Temp diff) - (Daily Range + 14)/2
BTU/YEAR = Sq Ft x U x Avg Temp. Diff. x Hrs

Wendes Mechanical Consulting Services Co. Form ENG 402

PEAK HEATING, COOLING AND CFM PER AREA

EXISTING ____ NEW ____

BUILDING ______________________ DATE ______________

______________ TOTAL SQ FT ______________ AVERAGE CFM PER TON ______________

AREA	SQ FT OF AREA	COOLING LOAD: SQ FT PER TONS	COOLING LOAD: TONS	HEATING LOAD: BTU PER SQ FT	HEATING LOAD: MBH	CFM SUPPLY: CFM PER SQ FT	CFM SUPPLY: CFM	DIRECT EXT. CFM
TOTAL								

Wendes Mechanical Consulting Services Co. Form ENG 403

Full Size Blank Testing and Balancing Audit Forms for Your Use (Continued)

Job ______________________ Date ______________

ENERGY CONSUMPTION HISTORY

PRIMARY HEAT

☐ Natural Gas Btu per cu. ft. ________
☐ Oil Btu per gal. ________
☐ Propane Btu per gal. ________
☐ Electricity Btu per KW, 3412
☐ Coal Btu per lb. ________

Fill in for a complete one year period

FUEL

Period	Days	Cubic Feet Gallons or LBS Consumed	BTU for Period*	Cost
Total For Year				

ELECTRIC

Period	Days	KW's Consumed	BTU for Period*	Cost
Total For Year				

*Multiply energy consumed by BTU per unit.
Example: 1000 cu. ft. gas x 1030 BTU/CU FT = 103,000 BTU

Wendes Engineering and Contracting Services Form AUD 302

Full Size Blank Testing and Balancing Audit Forms for Your Use (Continued)

Job ______________________ Date ______________

OUTSIDE WALL ENERGY EVALUATION

List all materials including air space
Enter R factor and total.

	Thick-ness Inches	R		
		Exist.	New 1	New 2
Interior Surface		.68	.68	.68
Surface				
Total R				
$\frac{1}{R_{total}}$ =	U			

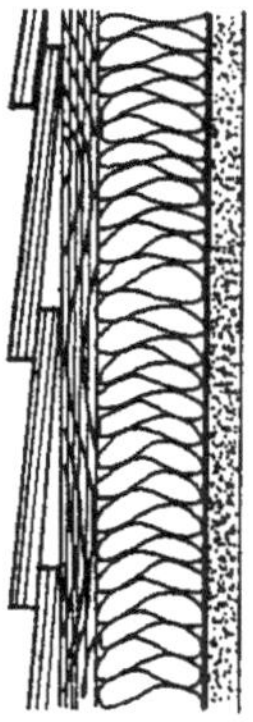

SQUARE FOOT AREA

EXISTING CONDUCTION HEAT LOSS PER YEAR

[______ x ______ x ______ x ______] ÷ ______ = [ ]

(U_x) (Net Area Sq Ft) (Avg Winter Temp Diff) (Total Winter Hours) Divide by: (Efficiency of Heating Equipment) (Heat Loss Mill. BTU Per Year)

PROPOSED ENERGY IMPROVEMENTS

Description ____________________________

New Heat Loss: (Same Formula as Above) New 1 New 2

[______ X ______ X ______ X ______] ÷ ______ = [ ]

[______ X ______ X ______ X ______] ÷ ______ = [ ]

Heat Loss Reduction Per Year — Subtract new heat loss from existing = [ ] [ ]

Fuel Reduction Per Year — Divide by BTU's per unit of fuel ☐Cu Ft ☐Therm ☐Gallon ☐KW → ÷ ______ [ ] ÷ ______ [ ]

Cost Savings Per Year — Multiply by ☐cost per unit of fuel or ☐cost per million BTU → X $______ per X $______ per [$] [$]

Wendes Engineering and Contracting Services Form A-3

Full Size Blank Testing and Balancing Audit Forms for Your Use (Continued)

Job ______________________________ Date ____________________

CEILING AND ROOF ENERGY EVALUATION

List all materials including air space
Enter R factor and total.

	Thickness Inches	R Exist.	R New 1	R New 2
Interior Surface		.68	.68	.68
Surface				
Total R				
$\frac{1}{R_{total}}$ =	U			

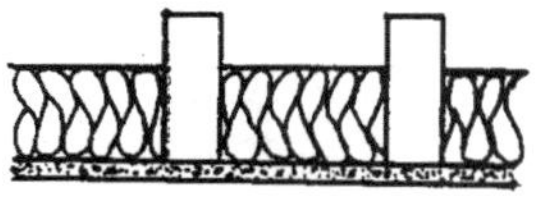

SQUARE FOOT AREA

EXISTING CONDUCTION HEAT LOSS PER YEAR

[______ x ______ x ______ x ______] ÷ ______ = [ ]

(U_x) (Net Area Sq Ft) (Avg Winter Temp Diff) (Total Winter Hours) (Divide by: Efficiency of Heating Equipment) (Heat Loss Mill. BTU Per Year)

PROPOSED ENERGY IMPROVEMENTS

Description ______________________________

New Heat Loss: (Same Formula as Above) *New₁* *New₂*

[______ X ______ X ______ X ______] ÷ ______ = [ ]

[______ X ______ X ______ X ______] ÷ ______ = [ ]

Heat Loss Reduction Per Year — Subtract new heat loss from existing = [ ] [ ]

Fuel Reduction Per Year — Divide by BTU's per unit of fuel ☐Cu Ft ☐Therm ☐Gallon ☐KW → ÷ ______ [ ] ÷ ______ [ ]

Cost Savings Per Year — Multiply by ☐cost per unit of fuel or ☐cost per million BTU → X $______ per X $______ per [$] [$]

Wendes Engineering and Contracting Services

Form AUD 314

Full Size Blank Testing and Balancing Audit Forms for Your Use (Continued)

WINDOWS AND DOORS

Job ______________________ Job No.________ Date ____________

SINGLE GLASS R=1 U=1		
Size	Quantity	Sq. Ft. Area
Total sq. ft.		

DOUBLE GLASS R=2 U=.5		
Size	Quantity	Sq. Ft. Area
Total sq. ft.		

Single glazed windows can be double glazed or covered with plastic. This will cut the heat loss in half.

If windows are triple glazed change R to 3, U=.33

Hollow core doors R=
Insulated or solid doors R=

OUTSIDE DOORS		
Size	Quantity	Sq. Ft. Area
Total sq. ft.		

EXISTING CONDUCTION HEAT LOSS PER YEAR

[________ x ________ x ________ x ________] ÷ ________ = [ ]

(U_x) (Net Area Sq Ft) (Avg Winter Temp Diff) (Total Winter Hours) Divide by: (Efficiency of Heating Equipment) (Heat Loss Mill. BTU Per Year)

Doors: [________ X ________ X ________ X ________] ÷ ________ = [ ]

PROPOSED ENERGY IMPROVEMENTS

Description ____

New Heat Loss: (Same Formula as Above) New 1 New 2

[________ X ________ X ________ X ________] ÷ ________ = [ ]

[________ X ________ X ________ X ________] ÷ ________ = [ ]

Heat Loss Reduction Per Year — Subtract new heat loss from existing = [ ] [ ]

Fuel Reduction Per Year — Divide by BTU's per unit of fuel ☐Cu Ft ☐Therm ☐Gallon ☐KW → ÷ ________ [ ] ÷ ________ [ ]

Cost Savings Per Year — Multiply by ☐cost per unit of fuel ☐cost per million BTU → \ $ ________ per \ $ ________ per [$] [$]

Wendes Engineering and Contracting Services

Full Size Blank Testing and Balancing Audit Forms for Your Use (Continued)

Job ______________________ Date ______________

COLD FLOOR ENERGY EVALUATION

List all materials including air space
Enter R factor and total.

	Thickness Inches	R Exist.	R New 1	R New 2
Interior Surface		.68	.68	.68
Surface				
Total R				
$\frac{1}{R_{total}}$ =	U			

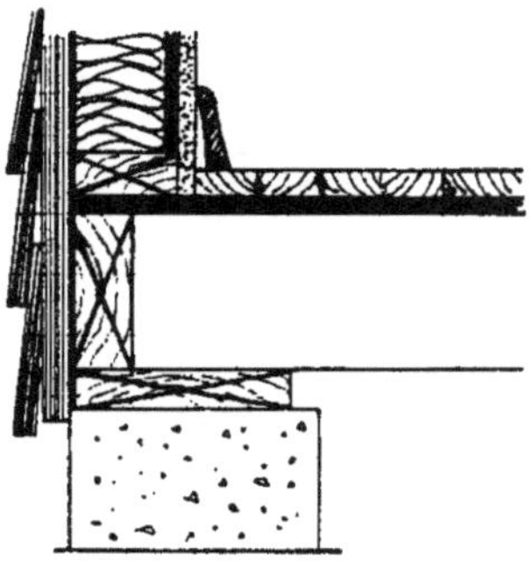

SQUARE FOOT AREA

EXISTING CONDUCTION HEAT LOSS PER YEAR

[______ x ______ x ______ x ______] ÷ ______ = [ ]

(U_x) (Net Area Sq Ft) (Avg Winter Temp Diff) (Total Winter Hours) Divide by: (Efficiency of Heating Equipment) (Heat Loss Mill. BTU Per Year)

PROPOSED ENERGY IMPROVEMENTS

Description ______________________________

New Heat Loss: (Same Formula as Above) *New 1* *New 2*

[______ X ______ X ______ X ______] ÷ ______ = [ ]

[______ X ______ X ______ X ______] ÷ ______ = [ ]

Heat Loss Reduction Per Year — Subtract new heat loss from existing = [ ] [ ]

Fuel Reduction Per Year — Divide by BTU's per unit of fuel ☐Cu Ft ☐Therm ☐Gallon ☐KW → ÷ ______ [ ] ÷ ______ [ ]

Cost Savings Per Year — Multiply by ☐cost per unit of fuel or ☐cost per million BTU → X $ ______ per X $ ______ per [$] [$]

Wendes Engineering and Contracting Services

Form AUD 31?

Full Size Blank Testing and Balancing Audit Forms for Your Use (Continued)

INDOOR AIR QUALITY REPORT

Job ______________________________ Date ______________

Location ______________________________ Longitude ______________

VENTILATION

Building Under Pressure: ☐ Positve ☐ Negative ☐ Neutral

Building Pressure Readings, Inches S.P.: Inside ________ Outside ________ No. of Occupants: ________

	Rated CFM	**Actual** CFM
Total Building Outside Air		
Total Building Exhaust Air		
Total Building Return Air		
Total Building Supply Air		
Infiltration		
Exfiltration		

CFM Outside Air Per Person ______________

Air Changes Per Hour ______________

CARBON DIOXIDE CO_2 READINGS

____

____

CARBON MONOXIDE CO READINGS

____

____

RELATIVE HUMIDITY RH READINGS

____

____

INDOOR CONTAMINANTS

____

____

OUTDOOR CONTAMINANTS

____

____

BIOLOGICAL CONTAMINANTS

____

____

INDUSTRIAL INDOOR CONTAMINANTS

____

PARTICULATES

BUILDING MATERIALS

INADEQUATE VENTILATION
- Inadequate *Outside Air*
- Improperly *Controlled* Outside Air
- Inadequate *Exhausts*
  - Toilet, Kitchen, Return Air, Fume, etc.
- *Tight* Building
- *Short Circuiting* of Air in Spaces
- Improperly *Pressurized* Building
- Inadequate *Filtering* of Air

INDOOR GENERATED CONTAMINANTS
- Tobacco Smoke
- Gas Leaks
- Freon Leaks
- Aerosols, Cleaners
  - Hairsprays, Cleaning Sprays, Disinfectants
- Pesticides
- Fumes, Chemical
- Nitrogen Dioxide
- Products of Combustion, Carbon Monoxide
  - Holes in Heat Exchanger and Flues
  - Clogged Flue or Chimney
  - Engine Exhausts, car, trucks
- Copy Machines
- Lasers
- Habachis and Charcoal Broilers

BUILDING MATERIALS, FABRICS, FURNISHINGS
- Carpets
- Sheets and Blankets
- Carpet Adhesives
- Furniture
- Fabrics in Furniture and Drapes, etc.
- Wood, Plywood
- Insulation
- Paneling, Particle Board
- Plastics, Laminate
- Asbestos

INDOOR BIOLOGICAL CONTAMINATION
- Standing Water
- Cooling Coils, Drain Pans
- Mildew Spores
- Carbon Dioxide CO_2 from People
- Pollen
- Mold
- Fungi
- Dust Mites
- House Dust
- Animal Dander
- Bacteria and Viruses
- Humidifiers (not evaporative type)

OUTDOOR CONTAMINANTS
- Due to Infiltration
- Radon
- Soil Gas
- Methane
- Pesticides
- Auto Pollution
- Exhaust Stacks
- Due to Outside Air Intake Drawing in Contaminated Industrial Process

INDUSTRIAL INDOOR CONTAMINANTS
- Paint
- Chemicals
- Printers
- Particulates, small solid or liquid particles such as dusts, powders, liquid droplets and mists.
  - Examples: flyash and asbestos dust.
- Gas Pollutants, fluids without form that occupy space rather than uniformly such as carbon monoxide or chloroform.
- Fumes, irritating smoke, vapor or gas.

Indoor Air Quality Forms (Continued)

BUILDING VENTILATION AND PRESSURE REPORT

Job ______________________________ Date ______________

Location ______________________________ Longitude ______________

Size of Building, Sq Ft ______________________________ No. of Floors ______ No. of Bldgs. ______

Building Under Pressure: ☐ Negative ☐ Positive ☐ Neutral

Pressure Readings: Inside _____ Outside _____ Floor _____ Walls: N______ S_______ E______ W______

Time: ☐ Winter ☐ Spring ☐ Summer ☐ Fall ☐ Daytime ☐ Night

System Tag	Type System	Location In Building	☐ Outside A ir ☐Recirc Air		☐ Exhaust Air[1] ☐Supply Fan[2]			
			Maximum CFM		Minimum CFM		Random CFM	
			Rated	Actual	Rated	Actual	Rated	Actual

Door, Window, Building Opening Conditions ______________________________

____

Stack Effect ______________________________

____

Wind Effect ______________________________

____

Combustion Effect ______________________________

____

Remarks ______________________________

____

____

____

Indoor Air Quality Forms (Continued)

HEAT LOSS CALCULATION

PEAK PER HR ____ SEASONAL ____ EXISTING ____ NEW ____

Building ______________________ Date ______________

Location ______________________ Latitude ______________

Type Building ______________ Stories ________ When Built ________

Sq Ft Area ______________ Cubic Ft of Space ______________

Calculation for Whole Building ____ For Partial Area ______________

Budget Load: Sq Ft ________ x BTUYR/Sq Ft ________ = ________

Outside Design: DB ________ WB ________ Avg OA Temp. ________ Winter Hours ________

Inside Design: DB(day) ________ RH ________ DB(night) ________

ITEM	DIMENSIONS	SQ FT	U	TEMP DIFF	BTU: Per Hour Seasonal	SEASONAL HOURS
ROOF OR CEILING						
FLOOR						
GLASS						
DOORS						
WALLS						
COLD INSIDE WALLS						
VENTILATION						
DUCT LOSSES						
GROSS TOTAL HEAT LOSS						
HEAT GAINS; LITES						
PEOPLE	No. People =					
OFF. EQUIPMENT	HP =					
COMPUTERS	KW =					
TOTAL INTERNAL GAINS						
NET TOTAL BUILDING HEAT LOSS						
INPUT TO HEATING EQUIPMENT, efficiency =						

BTUH = Sq Ft x U x Temp. Diff. BTU/YEAR = Sq Ft x U x Avg Temp. Diff. x Winter Hours
* BTU/SQ FT/YEAR

Remarks ______________________________

Wendes Mechanical Consulting Services Co. Form ENG 401

COOLING LOAD CALCULATION

PEAK PER HR ---- SEASONAL ---- EXISTING ---- NEW ----

Building ____________________ Date ____________________

Location ____________________ Latitude ________ Peak Load, HR, MO: ____________

Type Building ____________________ Stories ____________ When Built ____________________

Sq Ft Area ____________________ Cubic Ft of Space ____________________

X Calculation for Whole Building ------ For Partial Area ____________________

Building Load: Sq Ft ______________ x Sq Ft/Ton ______________ = ____________________

Outside Design: DB __________ WB ________ Avg OA Temp. ________ Summer Hours ____________________

Inside Design: DB(day) ______________ RH __________ DB(night) ____________

ITEM (Orient.)	DIMENSIONS	SQ FT	U or FACTOR*	TEMP DIFF	SENSIBLE BTUH	TONS	LATENT BTUH
ROOF OR CEILING							
GLASS							
(Solar)							
GLASS							
(Conduction)							
WALLS							
(Conduction)							
LIGHTING							
PEOPLE, NO.	(NO.)						
OFF. EQUIPMENT	MOTORS (HP)						
COMPUTERS	(KW)						
KITCHEN							
VENTIL. AIR, Sens.							
VENTIL. AIR, Lat.							
TOTAL COOLING LOAD,			SENSIBLE				
			LATENT				
			GRAND TOTAL				

BTUH = Sq Ft x U x Temp. Diff.
* U, CLTD, CLF, KW, HP, WD
CLTD = (Temp diff) - (Daily Range + 14)/2
BTU/YEAR = Sq Ft x U x Avg Temp. Diff. x Hrs

Wendes Mechanical Consulting Services Co. Form ENG 402

Engineering Calculation Forms for Your Use (Continued)

PEAK HEATING, COOLING AND CFM PER AREA

EXISTING ----- NEW -----

BUILDING -- DATE --------------------

-------------------------- TOTAL SQ FT ------------------ AVERAGE CFM PER TON --------------

AREA	SQ FT OF AREA	COOLING LOAD: SQ FT PER TONS	COOLING LOAD: TONS	HEATING LOAD: BTU PER SQ FT	HEATING LOAD: MBH	CFM SUPPLY: CFM PER SQ FT	CFM SUPPLY: CFM	DIRECT EXT. CFM
TOTAL								

Wendes Mechanical Consulting Services Co. Form ENG 403

Engineering Calculation Forms for Your Use (Continued)

HVAC ESTIMATE

Date ____________________

Job ____

Bid Date ____________________ Time ____________________

Place ____

Estimator ____

Architect ____________________ Phone ____________________

Engineer ____________________ Phone ____________________

JOB DESCRIPTION AND BUDGET COSTS

Job ______________________________ Date ______________

Location ______________________________ Distance ______________ Miles

Total Project Costs $ ______________ Volume of Building ______________ Cu Ft

Total Area ______________ Sq Ft, $Area_1$ ______________ Sq Ft $Area_2$ ______________ Sq Ft

BUDGET COSTS

	COST/SQ FT BLDG	TOTAL	COST/TON	TOTAL
Total HVAC	$	$	$	$
Sheet Metal	$	$	$	$
Piping	$	$	$	$
Equipment	$	$	$	$
Insulation	$	$	$	$
Temperature Control	$	$	$	$
Electric	$	$	$	$

DESIGN LOADS

	$Area_1$		$Area_2$	
	Factor	Total	Factor	Total
Cooling	Sq Ft/Ton	Tons	Sq Ft/Ton	Tons
Cooling	BTU/Sq Ft	BTU	BTU/Sq Ft	BTU
Heating	BTU/Sq Ft	BTU	BTU/Sq Ft	BTU
Supply Air	CFM/Sq Ft	CFM	CFM/Sq Ft	CFM
Duct Weight	LBS/Sq Ft	LBS	LBS/Sq Ft	LBS

SYSTEM DESCRIPTION

Heating
Cooling
☐ SZ ☐ MZ ☐ Constant Volume ☐ VAV
Duct Pressure ☐ LP ☐ MP ☐ HP
Return Air Method ☐ Duct ☐ Ceil. Plenum
Type Outlets
Type Perimeter Heat ☐ Air ☐ Baseboard
Temperature Control

KEY PLAN

No. of Buildings Stories

CONSTRUCTION: Glass ______________ Gross Area ________ Sq Ft; U ________

Exterior Walls ______________ Gross Area ________ Sq Ft; U ________

Roof ______________ Gross Area ________ Sq Ft; U ________

Remarks ______________________________

Wendes Engineering and Contracting Services 8/80 Form EST 101

HVAC Cost Estimating Forms (Continued)

SPECIAL LABOR	CORRECTION FACTOR AREAS
☐ Cartage	
☐ Shop Drawings	
☐ Field Sketching	
☐ Testing and Balancing	ALTERNATES
☐ Leak Testing	
☐ Service	
☐ Temporary Heat	
☐	ADDENDUMS
☐	
☐ Engineering	
☐ Sleeves	
☐ Removal	CONSTRUCTION SCHEDULE
☐ Property Protection	
☐ Cut Openings and Patch	
SUB CONTRACTORS AND RENTALS	
☐ Insulation	BIDDING PLAN AND SPEC NO.'S
☐ Temperature Control	
☐ Piping	
☐ Electrical	
☐	NOT INCLUDED
☐ Concrete Pads	
☐ Cranes	
☐	
☐ Testing and Balancing	
☐ Excavating and Backfilling	
☐ Cutting and Patching	
☐	
END OF BID ITEMS	
☐ Sales Tax	
☐ Permits	
☐ Bid Performance	
☐	
☐ Travel Pay	
☐ Room and Board	
☐ Wage Increases	
☐ Material Price Increases	
☐ Premium Time	
☐ Contingencies	

Wendes Engineering and Contracting Services

HVAC Cost Estimating Forms (Continued)

BID CHECK OFF SHEET

☐ Total HVAC ☐ Sheet Metal ☐ Piping

Job ______________________________ Due Date ______________

Location ______________________________ Distance ______________

Project Costs $ __________ Size, Sq Ft __________ Stories __________

Budget Estimate ______________________________ Buildings __________

EQUIPMENT	☐ DUCTWORK ☐ PIPING
	SPECIALTIES
	INSULATION

Wendes Engineering and Contracting Services 7/80 Form EST 102

HVAC Cost Estimating Forms (Continued)

QUOTATION CALL LIST AND PRICE COMPARISON

Job ______________________ Due Date ______________________

Location ______________________ Estimator ______________________

Architect/Engineer ______________________ Date ______________________

ITEM	MANUFACTURER	SUPPLIER	PHONE	QTY	QUOTED PRICE	

Wendes Engineering and Contracting Services

7/80 Form EST 104

HVAC Cost Estimating Forms (Continued)

QUANTITY TAKEOFF SHEET

Page ________

Date ________

Job ________________________________

DESCRIPTION

Drwg Flr or Sys	Size	Description	Quantity	Total							

Wendes Engineering and Contracting Services   10/80 Form EST 105

HVAC Cost Estimating Forms (Continued)

ESTIMATE SUMMARY & EXTENSION SHEET

Date ____________

Page ____________

Job ____________

		MATERIAL COST		SHOP		FIELD	
		UNIT	TOTAL	UNIT	TOTAL	UNIT	TOTAL

Wendes Engineering and Contracting Services

8/80 Form EST 106

HVAC Cost Estimating Forms (Continued)

Date ____________________

TELEPHONE QUOTATION

Job	
Supplier	Phone
	By

QTY	MFGR	DESCRIPTION	ACCESS-ORIES	AMOUNT	
				Each	Total
		Grand Total			

NOT INCLUDED					

Meets plans and specs ☐	Taxes included ☐
Addendums included ☐	Freight included ☐
Type materials correct ☐	Lead time required
	Price good for ________ days

Quote Received By: ____________________

Wendes Engineering and Contracting Services 9/80 Form No. EST 107

HVAC Cost Estimating Forms (Continued)

BID RECAP AND MARKUP SHEET

Job ______________________ Due Date ______________

Location ______________________ Estimator ______________

	HOURS	WAGE RATE	COST			
Shop Labor			$			
Field Labor			$			
Wage Increase Shop			$			
Wage Increase Field			$			
Overtime			$			
Travel Costs			$			
			$			
			$			
			TOTAL LABOR		$	
Raw Materials			$			
Equipment			$			
			$			
Sales Tax			$			
			TOTAL MATERIAL AND EQUIPMENT		$	
Subcontracts			$			
			$			
			$			
			$			
			$			
			$			
			TOTAL SUBCONTRACTS		$	
			TOTAL DIRECT COSTS		$	
Overhead On Labor		%	$			
Overhead On Material and Equipment:		%	$			
Overhead On Subcontractors:		%	$			
			TOTAL OVERHEAD		$	
(% of Total Direct Costs) (% of Sales)						
			TOTAL DIRECT AND INDIRECT COSTS		$	
Profit: % of Total Costs			$			
Performance Bond: % of Total Bid			$			
Financing Costs: Amount $; %			$			
Total O&P Markup $, Percent of Sales %					$	
			TOTAL BID PRICE		$	

Budget Check: ______________________

Wendes Engineering and Contracting Services 7/80 Form EST 108

HVAC Cost Estimating Forms (Continued)

BIDDING RECORD

Job ______________________ Due Date ______________

Location ______________________ Time ______________

BID SUBMITTED TO:

Company	Name	Phone	Amount	Remarks

INCLUSIONS

____
____
____
____
____
____
____

EXCLUSIONS

____
____
____
____
____
____
____

ADDENDUMS

____
____
____

ALTERNATES

____
____
____

Drawings Included ______________________

Specifications Included ______________________

Remarks ______________________

Wendes Engineering and Contracting Services 1/81 FORM EST 109

HVAC Cost Estimating Forms (Continued)

CALCULATING LABOR COSTS PER HOUR

Location ______________________ Date ______________

☐Union; Local No. __________ Contract Expiration __________ Non Union __________

FRINGE BENEFITS PER HOUR

	Journeyman	Foreman	General Foreman	Other
Base Rate	$	$	$	$
Welfare (Medical) (%)				
Pension (%)				
Apprentice Fund				
National Training Fund				
Vacation Savings				
Industry Fund				
TOTAL BENEFITS %	$	$	$	$
TOTAL WITH BASE	$	$	$	$

PAYROLL TAXES AND INSURANCE

F.I.C.A. %				
Workman's Comp. %				
Federal Unemployment %				
State Unemployment %				
Liability Insurance %				
Property Insurance %				
Association Due %				
TOTAL TAXES & INS. %				
TOTAL BASE, BENEFITS, TAXES, INS				

COST PER POUND BREAKDOWN

	LBS/HR	COST/LB*	LBS/HR	COST/LB*
Material				
Shop Labor				
Field labor				
Shop Drawings				
Cartage				
TOTAL DIRECT COSTS		$		$
Indirect Overhead %				
TOTAL COSTS				
Profit %				
TOTAL SELL		$		$

*Based on total labor cost above $

HVAC Cost Estimating Forms (Continued)

COMPANY MARKUP CALCULATION SHEET

Date ______________________

Company ______________________ Period ______________________

1. Anticipated Sales For Year $ __________ Percent of Sales

2. Total Indirect Overhead and Administration Costs for Year $ __________ ______

   Profit Desired ______ % $ __________ ______

   Total Anticipated Direct Costs for Year (Material & Labor) $ __________ ______

   Breakdown: Labor (Includes fringes, payroll taxes, ins.) $ __________

   Material and Equipment $ __________

   Subs $ __________

3. SINGLE MARKUP NEEDED ON TOTAL DIRECT COSTS

   Percent For Overhead Only: $ Overhead Costs / $ Direct Costs = $ ______ / $ ______ = ______ %

   Percent For Overhead and Profit: $ Overhead & Profit / $ Direct Costs = $ ______ / $ ______ = ______ %

4. SIMPLIFIED DUAL MARKUP FOR OVERHEAD — Amount of Markup For Year

   Markup on Materials and Equipment ___% x $ ______ = $ ______

   Markup on Subs ___% x $ ______ = $ ______

   Total Overhead on Mat. & Subs $ ______

   Percent Markup on Labor for Overhead = ($Total Ovhd) - ($Matl. & Sub Ovhd) / (Labor Costs) =

   $(______) - ($ ______) / $(______) = $ ______ / $ ______ = (______) %

5. TOTAL SELLING COST OF LABOR PER HR WITH DUAL MARKUP SYSTEM

   Wages per hr (incl. fringes, ins. and taxes) $ ______

   $Wages x percent overhead markup on labor $ ______

   Profit ___ % $ ______

   Total $ ______

Mendes Engineering and Contracting Services 1/84 FORM EST 112

HVAC Cost Estimating Forms (Continued)

PER PIECE DUCT TAKEOFF SHEET

Job__________ Drawing________ System_______ ☐ Lining________

Type Duct: ☐Galv, ☐LP ☐HP, ☐Other__________ ☐ Insulation________

DUCT SIZE	TYPE DUCT	EQUIVALENT LINEAR FEET PER PIECE	TOT LF	WEIGHT		QTY	SHOP LABOR		FIELD LABOR	
				LBS LF	Total		Hrs /Pc	Total	Hrs /Pc	Total

Wendes Engineering and Contracting Services

Form No. EST 120

HVAC Cost Estimating Forms (Continued)

PER POUND DUCT TAKEOFF SHEET

Job____________________ Drawing________ System________ ☐ Lining________

Type Duct: ☐ Galv, ☐ LP ☐ HP, ☐ Other__________ Duct Elevat.____ ☐ Insulation______

DUCT SIZE	SQ. FT PER LINEAR FEET	LINEAR FEET	TOTAL LINEAR FEET	LBS PER LF with WASTE	0-12 26ga	13-30 24ga	31-54 22ga	55-84 20ga	85 up 18ga	T V	V D	F C	F D	

Wendes Engineering and Contracting Services 9/80 Form No. EST 122

HVAC Cost Estimating Forms (Continued)

PIPING TAKEOFF SHEET

Job ______ Drawing ______ System ______ Factors ______

Type Pipe ______ Connections ______ ☐ Insulation ______

DIAMETER	ITEM	QUANTITY OR LINEAR FEET	TOTAL	MATERIAL COST		ERECTION LABOR		
				Unit	Total	Equiv. Joints	Unit	Total

Wendes Engineering and Contracting Services

Form No. EST 130

HVAC Cost Estimating Forms (Continued)

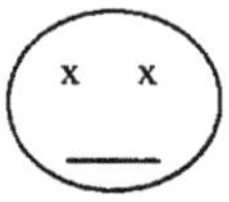

PROBLEM WORK SHEET

Hitting Home Runs by Wendes

Date: ____________________ Page _________of _________

PROBLEM.......(brief statement):

OVERALL PROBLEM:

General Area or Category	PROBLEMS (main problems and sub-problems)		Possible Key Causes or Options
	General Description	Specifics	

Assessment of Problem: Importance: urgency, size, complexity. Type problem. "State" problem is in. Risks of problem.

CAUSE WORK SHEET

Hitting Home Runs by Wendes

Date: ____________________ Page_________ of ___________

PROBLEM.........(brief statement):

General Description, Symptoms, etc.:

General Area or Category	POSSIBLE CAUSES		1. Chances of being Cause, 2. Possible Degree of Effect
	General Description	Specifics	

Information, Tests, Research Needed: Priorities of testing.:

OBJECTIVES WORK SHEET

Hitting Home Runs by Wendes

Date: ____________________ Page ________ of ________

PROBLEM:

OVERALL GOAL:

General Category of Consideration	OBJECTIVES		Importance Ratings, Evaluations
	General Description	Specifics	

Remarks:

OPTIONS WORK SHEET

Hitting Home Runs by Wendes

Date____________________ Page________ of__________

PROBLEM.........(brief statement):

Information and Planning Needed:

General Area or Category	OPTIONS (Possible solutions)		Effects, Risks, Odds, Ratings
	General Description	Specifics	

Remarks

(Form can be used for outline list of all the options, and for evaluating each option separately)

i/p

INFORMATION WORK SHEET

- *Information, Research, Tests, Consulting Needed*
- *Relationships, Diagrams, Planning and Scheduling*

Hitting Home Runs, by Wendes

Date: ____________ Page ______ of ______

PROBLEM:

Remarks:

Appendix B

Charts and Formulas

ABBREVIATIONS AND SYMBOLS

A	Area, Square Feet
A	Amps
act	Actual
bhp	Break Horsepower
Btu	British Thermal Unit
Btuh	British Thermal Unit per Hour
cfm	Cubic Feet per Minute
Cu	Cubic
Cu ft	Cubic Feet
CHW	Chilled Water
CHWS	Chilled Water Supply
CHWR	Chilled Water Return
D	Difference, Delta
db	Dry Bulb Temperature
DIA, dia	Diameter
DP, ΔP	Difference in Pressure
Dt, Δt	Difference in Temperature
E	Voltage
eff	Efficiency
F	Fahrenheit
fpm	Feet per Minute
ft	Feet
GPH	Gallons per Hour
gpm	Gallons per Minute
H	Head
hp	High Pressure System
Hs	Elevation Head
HW	Hot Water
HWS	Hot Water Supply
HWR	Hot Water Return
I	Amperes
kVa	Kilovolt-Amperes
kW	Kilowatts
LP	Low Pressure System
MP	Medium Pressure System
MZ	Multi-Zone
NPSH	Net Positive Suction Head
Ns	Specific Speed
OA	Outside Air
P	Pressure
Pa	Atmospheric or Absolute Pressure
PF	Power Factor
Pvp	Vapor Pressure
R	Rankine Temperature
R	Resistance, Ohms
RA	Return Air
RH	Relative Humidity
rpm	Revolutions per Minute
SP	Static Pressure, Inches Water Gauge
SP gr	Specific Gravity
sq. ft	Square Feet
Sz	Single Zone System
t	Temperature, degrees Fahrenheit or Celsius
T	Absolute Temperature, 460 + Degrees Fahrenheit
TP	Total Pressure, Inches Water
V	Velocity, Feet per Minute
V	Volts
VP	Velocity Pressure, Inches Water Gauge
wb	Wet Bulb Temperature
wg	Water Gauge

ASSOCIATION ABBREVIATIONS

ACCA	Air Condition Contractors Association of America
ADC	Air Diffusion Council
ADI	Air Distribution Institute
AMCA	Air Moving and Control Association, Inc.
ARI	Air Conditioning and Refrigeration Institute
ASHRAE	American Society of Heating, Refrigeration and Air Conditioning Engineers, Inc.
ASME	American Society of Mechanical Engineers
MCAA	Mechanical Contractors Association of America, Inc.
NAPHCC	National Association of Plumbing-Heating-Cooling Contractors
NEMA	National Electrical Manufacturers Association
NSPE	National Society of Professional Engineers
SMACNA	Sheet metal and Air Conditioning Contractors National Association, Inc.

AIR FLOW AND AIR PRESSURE FORMULAS

Air Flow Formula

Used to find volume of air flowing through ductwork, outlets, inlets, hoods, etc.

Basic Formula $cfm = A \times V$

where:

cfm	=	cubic fpm
A	=	area in sq. ft
V	=	velocity in fpm
A_k	=	factor used with outlets; actual unobstructed air flow area

Velocity unknown:

$$V = \frac{CFM}{A}$$

Area unknown:

$$A = \frac{CFM}{V}$$

Total Pressure Formula

Measure of total pressure energy in air at any particular point in an air distribution system.

AIR FLOW AND AIR PRESSURE FORMULAS (CONT'D.)

TP = VP + SP

rearranged:

VP TP - SP

SP TP - VP

where:

TP = total pressure inches W.G.

VP = velocity pressure, inches W.B.

SP = static pressure, inches W.G.

CONVERTING VELOCITY PRESSURE INTO FPM

Standard Air, 075 lb/cu ft

fpm = 4005 × fV-1P

rearranged:

$$VP = \left(\frac{fpm}{4005}\right)^2$$

Non Standard Air

$$fpm = 1096 \times \sqrt{\frac{VP}{density}}$$

where:

VP = velocity pressure, inches W.G.

density = lb/cu ft

fpm = feet per minute

CHANGING FAN CFMs AND DRIVES

Fan Law No. 1

$$\text{RPM new} = \text{RPM old} \times \left(\frac{\text{CFM new}}{\text{CFM old}}\right)$$

$$\text{SP new} = \text{SP old} \times \left(\frac{\text{CFM new}}{\text{CFM old}}\right)^2$$

$$\text{BHP new} = \text{BHP old} \times \left(\frac{\text{CFM new}}{\text{CFM old}}\right)^3$$

where:

rpm = revolutions per minute
cfm = cubic feet per minute
SP = fan static pressure, inches W.G.
bhp = brake horsepower

BHP Formulas

$$\underset{\text{(3 phase)}}{\text{BHP actual}} = \frac{1.73 \times \text{amps} \times \text{volts} \times \text{eff.} \times \text{power factor}}{746}$$

$$\underset{\text{(rule of thumb)}}{\text{BHP actual}} = \frac{\text{(nameplate)}}{\text{horsepower}} \times \left(\frac{\text{amps act.}}{\text{amps rated}}\right) \times \left(\frac{\text{volts act.}}{\text{volts rated}}\right)$$

where:

bhp = brake horsepower
eff = efficiency

Sheave/RPM Ratios & Belt Lengths

$$\frac{\text{RPM motor}}{\text{RPM fan}}\left(\begin{array}{c}\text{Speed}\\ \text{Ratio}\end{array}\right) = \frac{\text{DIA fan sheave}}{\text{DIA motor sheave}}\left(\begin{array}{c}\text{Diameter}\\ \text{Ratio}\end{array}\right)$$

$$\text{DIA fan sheave} = \text{DIA motor sheave} \times \left(\frac{\text{RPM motor}}{\text{RPM fan}}\right)$$

$$\text{DIA motor sheave} = \text{DIA fan sheave} \times \left(\frac{\text{RPM fan}}{\text{RPM motor}}\right)$$

$$\text{Belt Length} = 2c + [1.57 \times (D + d)] + \frac{(D \pm d)^2}{4c}$$

where:

C = center to center distance of shaft
D = large sheave diameter
d = small sheave diameter

AIR HEAT TRANSFER FORMULAS

Sensible

$$\text{Btuh} = \text{cfm} \times \text{temp change} \times 1.08$$

Rearranged:

$$\text{CFM} = \frac{\text{BtuH (Sensible)}}{1.08 \times \text{temp change}}$$

Rearranged:

$$\text{Temp Change} = \frac{\text{BtuH (Sensible)}}{\text{CFM} \times 1.08}$$

Latent

$$\text{Btuh} = 4840 \times \text{cfm} \times \text{WD}$$

Total Latent and Sensible

$$\text{Btuh} = 4.5 \times \text{cfm} \times \text{HD}$$

where:

Btuh	=	British thermal units per hour
T	=	Temperature, F
cfm	=	Cubic feet per minute
EFF	=	Efficiency

AIR HEAT TRANSFER FORMULAS (CONT'D.)

HD = Difference in enthalpy
WD = Difference in humidity ratio (lb water/lb dry air)

AIR FLOW FOR FURNACES

Gas Furnace

$$\text{CFM} = \frac{\text{Heat value of gas (Btu/cu ft)} \times \text{cu ft/hr} \times 0.75}{1.08 \times \text{Temp Rise *}}$$

Oil Furnace

$$\text{CFM} = \frac{\text{Heat value of oil (Btu/gal)} \times \text{gal/hr} \times 0.70}{1.08 \times \text{Temp Rise *}}$$

Electric Furnace

$$\text{CFM} = \frac{\text{Volts} \times \text{Amps} \times 3.4}{1.08 \times \text{Temp Rise *}}$$

where:
Btuh = British thermal units per hour
RH = relative humidity, percent
T = temperature, IF
cfm = cubic feet per minute
*Difference between supply air and return air temperatures

CHANGES IN STATE OF AIR FORMULAS

For changes of state in volume, temperature and pressure of air and other gases.

New volume of air pressure remains constant and temperature changes

$$V\text{ new} = V\text{ orig} \times = \left(\frac{T\text{ new}}{T\text{ orig}}\right)$$

New volume when temperature remains constant and pressure changes

$$V\text{ new} = V\text{ orig} \times = \left(\frac{P\text{ orig}}{P\text{ new}}\right)$$

New pressure when volume is constant and temperature changes

$$P\text{ new} = P\text{ old} \times = \left(\frac{T\text{ new}}{T\text{ old}}\right)$$

Derived from gas laws:

$$\left(\frac{PV}{T}\right)_{\text{orig state}} = \left(\frac{PV}{T}\right)_{\text{new state}}$$

where:

V = volume in cu ft
P = absolute pressure (atmospheric + gauge pressure)
T = absolute temperature (460 + T)

AIR DENSITY CORRECTION FACTORS
FOR DIFFERENT ALTITUDES AND TEMPERATURES

Air	Sea	Attitude (ft)									
Temp. °F	Level	1000	2000	3000	4000	5000	6000	-7000	8000	9000	10,000
–40°	1.26	1.22	1.17	1.13	1.09	1.05	1.01	0.97	0.93	0.90	0.87
0°	1.15	1.11	1.07	1.03	0.99	0.95	0.91	0.89	0.85	0.82	0.79
40°	1.06	1.02	0.99	0.95	0.92	0.88	0.85	0.82	0.79	0.76	0.73
70°	1.00	0.96	0.93	0.89	0.86	0.83	0.60	0.77	0.74	0.71	0.69
100°	0.95	.0.92	0.68	0.85	0.81	0.78	0.75	0.73	0.70	0.68	0.65
150°	0.87	0.84	0.81	0.78	0.75	0.72	0.69	0.67	0.65	0.62	0.60
200°	0.80	0.77	0.74	0.71	0.69	0.66	0.64	0.62	0.60	0.57	0.55
250°	0.75	0.72	0.70	0.67	0.64	0.62	0.60	0.58	0.56	0.58	0.51
300°	0.70	0.67	0.65	0.62	0.60	0.58	0.56	0.54	0.52	0.50	0.48
350°	0.65	0.62	0.60	0.58	0.56	0.54	0.52	0.51	0.49	0.47	0.45
400°	0.62	0.60	0.57	0.55	0.53	0.51	0.49	0.48	0.46	0.44	0.42
450°	0.58	0.56	0.54	0.52	0.50	0.48	0.46	0.45	0.43	0.42	0.40
500°	0.55	0.53	0.51	0.49	0.47	0.45	0.44	0.43	0.41	0.39	0.38
550°	0.53	0.51	0.49	0.47	0.45	0.44	0.42	0.41	0.39	0.38	0.36
600°	0.50	0.48	0.46	0.45	0.43	0.41	0.40	0.39	0.37	0.35	0.34
700°	0.46	0.44	0.43	0.41	0.39	0.38	0.37	0.35	0.34	0.33	0.32
800°	0.42	0.40	0.39	0.37	0.36	0.35	0.33	0.32	0.31	0.30	0.29
900°	0.39	0.37	0.36	0.35	0.33	0.32	0.31	0.30	0.29	0.28	0.27
1000°	0.36	0.35	0.33	0.32	0.31	0.30	0.29	0.28	0.27	0.26	0.25

Standard Air Density, Sea Level, 70°F = 0.075 lb/cu ft

EXAMPLE:

Find actual density of air at sea level if the temperature is 600'F.

Actual Density = (0.75 lb/cu ft) × (0.52) = .04 lb/cu ft

Find the actual density of 70°F air at 5000 feet altitude.

Actual Density = (0.75 lb/cu ft) × (0.83) = .0622 lb/cu ft

CONVERTING VELOCITY PRESSURE INTO FEET PER MINUTE (STANDARD AIR)

Velocity Pressure	fpm
.001	127
.005	283
.010	400
.015	491
.020	566
.025	633
.030	694
.035	749
.040	801
.045	849
.050	896
.055	939
.060	981
.065	1020
.070	1060
.075	1097
.080	1133
.085	1167
.090	1201
.095	1234
.100	1266
.105	1298
.110	1328
.115	1358
.120	1387
.125	1416
.130	1444
.135	1471
.140	1498
.145	1525
.150	1551
.155	1577
.160	1602
.165	1627
.170	1651
.175	1675
.180	1699
.185	1723
.190	1746
.195	1768
.200	1791

Velocity Pressure	fpm
.21	1835
.22	1979
.23	1921
.24	1962
.25	2003
.26	2012
.27	2081
.21	2119
.29	2157
.30	2193
.31	2230
.32	2260
.33	2301
.34	2335
.35	2369
.36	2403
.37	2436
.38	2469
.39	2501
.40	2533
.41	2563
.42	2595
.43	2626
.44	2656
.45	2687
.46	2716
.47	2746
.48	2775
.49	2804
.50	2832
.51	2860
.52	2888
.53	2916
.54	2943
.55	2970
.56	2997
.57	3024
.58	3050
.59	3076
.60	3102

Velocity Pressure	fpm
.61	3127
.62	3153
.63	3179
.64	3204
.65	3229
.66	3254
.67	3279
.68	3303
.69	3327
.70	3351
.71	3375
.72	3398
.73	3422
.74	3445
.75	3468
.76	3491
.77	3514
.78	3537
.79	3560
.80	3582
.81	3604
.82	3625
.83	3657
.84	3669
.85	3690
.86	3709
.87	3729
.88	3758
.89	3779
.90	3800
.91	3821
.92	3842
.93	3863
.94	3894
.95	3904
.96	3924
.97	3945
.98	3965
.99	3985
1.00	4005

Velocity pressure in inches wg

Based on formula: $fpm = 4005\sqrt{VP}$

CONVERTING VELOCITY PRESSURE INTO FEET PER MINUTE

FOR VARIOUS TEMPERATURES

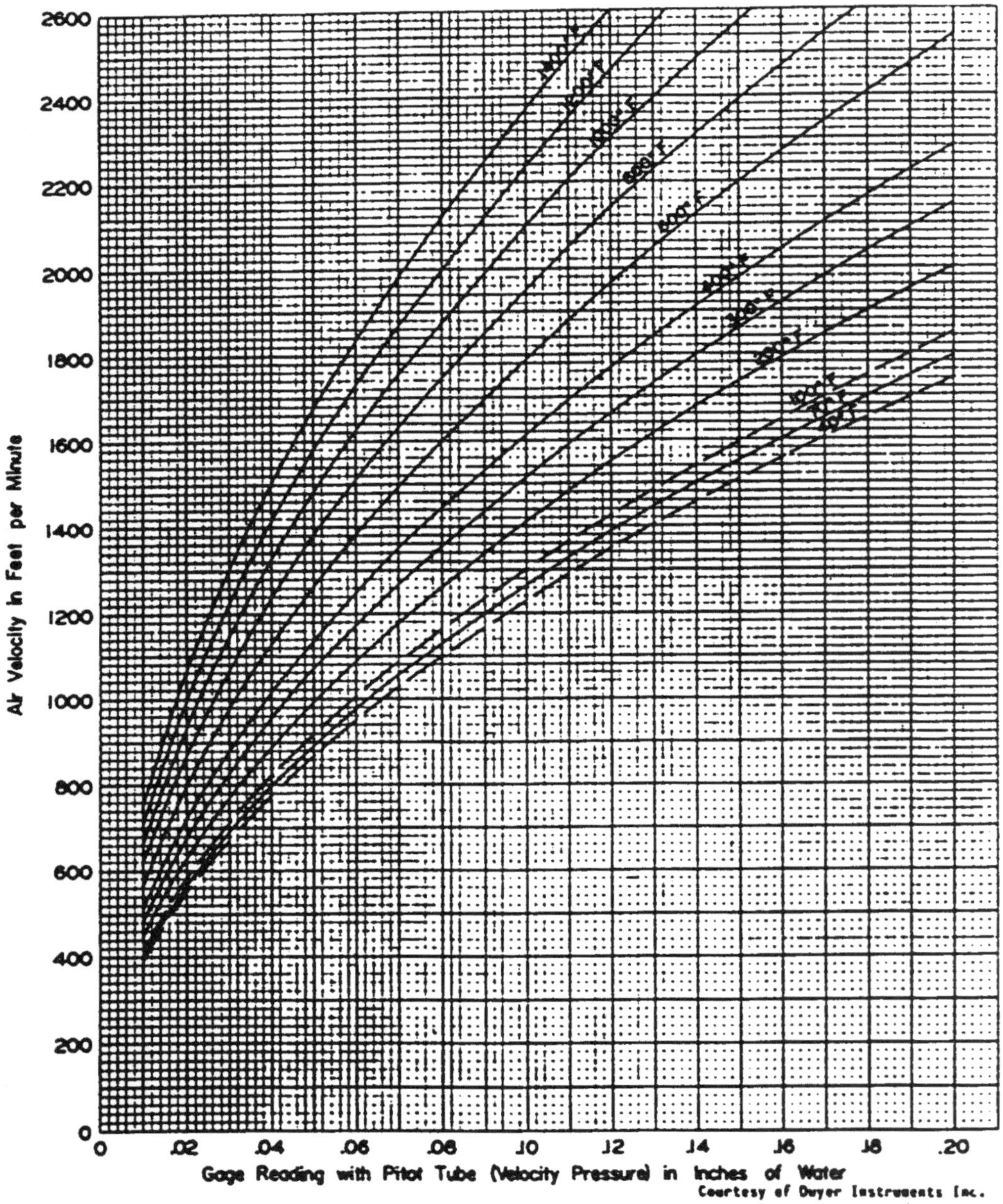

Courtesy of Dwyer Instruments Inc.

HYDRONIC FORMULAS

Converting PSI to Feet of Head:

$$\text{ft hd} = 2.31 \times \text{psi} \qquad \text{inches hd} = 27.2 \times \text{psi}$$

$$\text{psi} = .433 \times \text{ft. hd} \qquad \text{psi} = .036 \times \text{inches hd}$$

Water Heat Transfer Formulas

$$\text{Btuh} = \text{gpm} \times (\text{T in} - \text{T out}) \times 500$$

$$\text{GPM} = \frac{\text{BtuH}}{(\text{T in} \pm \text{T out}) \times 500}$$

Electrical Power Consumption of Water Pump

$$\text{BHP} = \frac{\text{GPM} \times \text{ft head}}{3960 \times \text{eff}}$$

Using System Component as Flow Measuring Device

$$\text{GPM actual} = \text{GPM design} \times \sqrt{\frac{\Delta \text{P actual}}{\Delta \text{P de sign}}}$$

$$\Delta \text{P actual} = \Delta \text{P design} = \times \left(\frac{\text{GPM actual}}{\text{GPM design}}\right)^2$$

Coil or Chiller GPM

$$\text{GPM} = \frac{\text{Tons} \times 24}{\text{T in} \pm \text{T out}}$$

Condenser GPM

$$\text{GPM} = \frac{\text{Tons} \times (\text{kW} \times .284)}{\text{T out} \pm \text{T in}}$$

where:

gpm = gallons per minute
ΔP = change in pressure across component
Btuh = British thermal units per hour
T = temperature, 'F
bhp = brake horsepower
kW = kilowatts

PSYCHROMETRIC CHART
(Sea Level)

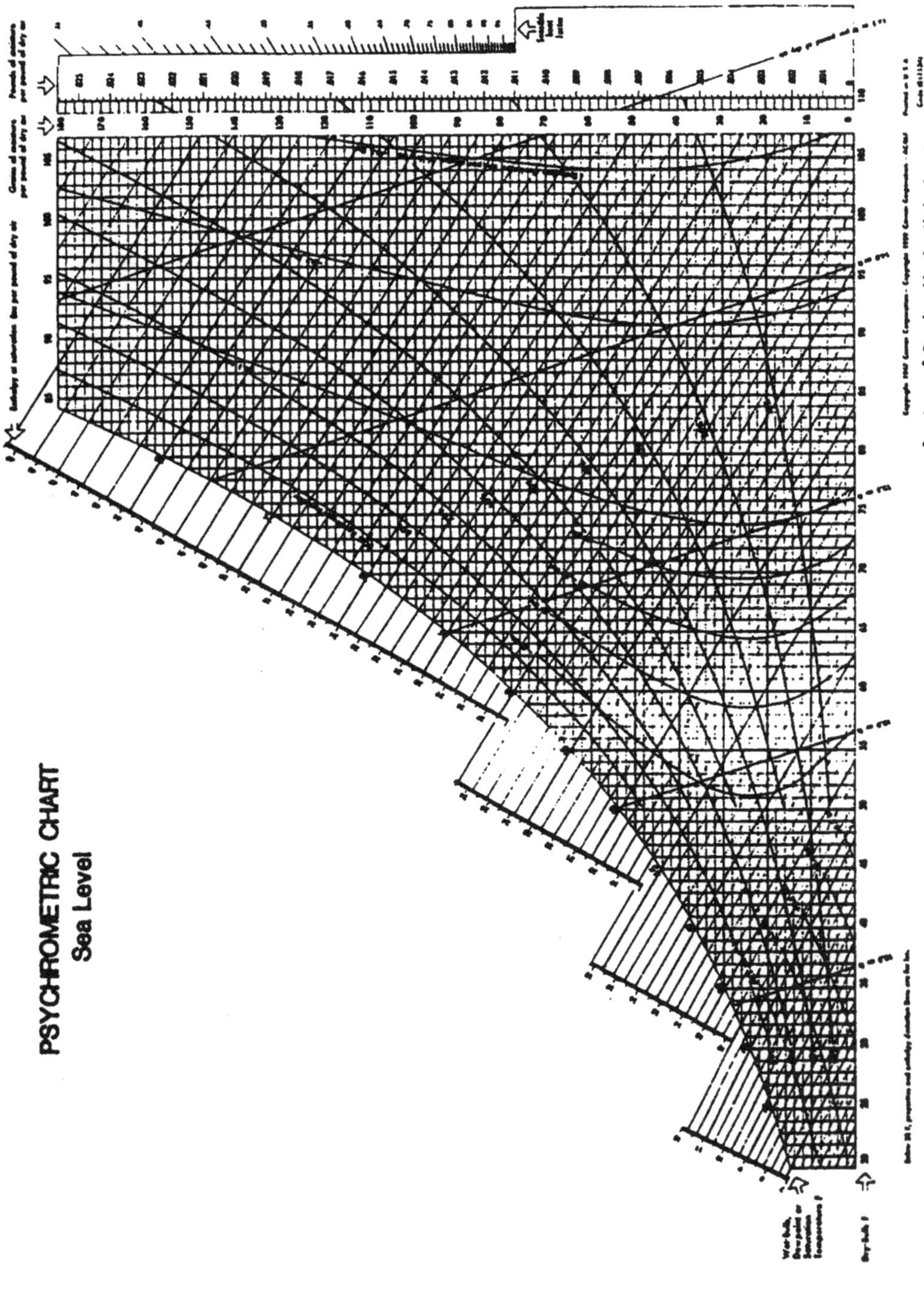

MOTOR AMP DRAWS, EFFICIENCIES, POWER FACTORS, STARTER SIZES

Approximate Values

Induction type motors, 1800 rpm, 3-phase, 60-cycle.

hp	Full Load Amps			Starter Size		Percent	Power
	115V	230V	460V	230V	460V	Efficiency	Factors
1/2	4.0	2.0	1.0	00	00	70	69.2
3/4	5.6	2.8	1.4	00	00	72	72.0
1	7.2	3.6	1.8	0	0	79	76.5
1-1/2	10.4	6.8	3.4	0	0	80	80.5
2	13.6	6.8	3.4	0	0	80	85.3
3	19.2	9.6	4.8	0	0	81	82.6
5	30.4	15.2	7.6	0	0	83	84.2
7-1/2		22	11	1	0	85	85.5
10		28	14	1	1	85	88.8
15		42	21	2	1	86	87.0
20		54	27	2	1	87	87.2
25		68	34	2	2	88	86.8
30		80	40	3	2	89	87.2
40		104	52	3	2	89	88.2
50		130	65	4	3	89	89.2
60		154	77	4	3	89	89.5
75		192	96	4	4	90	89.5
100		248	124	5	4	90	90.3
125		293	147			90	90.5
150		348	174			91	90.5
200		452	226			91	
250		568	214			91	
300		678	339			92	

Rule of Thumb on AMP Draws for Motors

- A 115V motor draws double the amps of a 230V motor.
- A 230V motor draws double the amps of a 460V motor.
- Single-phase motors draw double the amps of 3-phase motors.
- At 115 volts, a 3-phase motor draws about 5.2 amps per hp.
  At 230 volts, a 3-phase motor draws about 2.6 amps per hp.

CONVERSION FACTORS

MULTIPLY	BY	TO OBTAIN
Atmospheres (Std.)		
760 MM of Mercury		
at 32°F.	14,696	Lbs./sq. inch
Atmospheres	76.0	Cms. of mercury
Atmospheres	29.92	In. of mercury
Atmospheres	33.90	Feet of water
Atmospheres	1.0333	Kgs./sq. cm.
Atmospheres	14.70	Lbs./sq. inch
Atmospheres	1.058	Tons/sq. ft.
Brit. Therm. Units	0.2520	Kilogram-calories
Brit. Therm. Units	777.5	Foot-lbs.
Brit. Therm. Units	0.000393	Horse-power-hrs.
Brit. Therm. Units	0.293	Watt-hrs.
Btu/min.	12.96	Foot-lbs./sec.
Btu/min.	0.02356	Horse-power
Btu/min. Btu./min.	0.01757	Kilowatts
Calorie	17.57	Watts
Centimeters	0.003968	Btu
Centimeters	0.3937	Inches
Centimeters	0.03280	Feet
Centimeters	0.01	Meters
Centmtrs. of merc.	10	Millimeter.
Centmtrs. of merc.	0.01316	Atmospheres
Centmtrs. of merc.	0.4461	Feet of water
Centmtrs. of merc.	136.0	Kgs./sq. meter
Centmtrs. of merc.	27 85	Lbs./sq. ft.
Cubic feet	0.1934	Lbs./sq. inch
Cubic feet	2.832×10^6	Cubic cms.
Cubic feet	1728	Cubic inches
Cubic feet	0.02832	Cubic meters
Cubic feet	0.03704	Cubic yards
Cubic feet/minute	7.48052	Gallons U.S.
Cubic feet/minute	472.0	Cubic cms./sec.
Cubic foot water	0.1247	Gallons/sec.
Feet	62.4	Pounds @ 60°F.
Feet	30.48	Centimeters
Feet	12	Inches
Feet	0.3048	Meters
	1/3	Yards

MULTIPLY	BY	TO OBTAIN
Feet of water	0.02950	Atmospheres
Feet of water	0.8826	Inches of mercury
Feet of water	0.03048	Kgs./sq. cm.
Feet of water	62.43	Lbs./sq. ft.
Feet of water	0.4335	Lbs./sq. inch
Feet/min.	0.5080	Centimeters/sec.
Feet/min.	0.01667	Feet/sec.
Feet/min.	0.01829	Kilometers hr.
Feet/min.	0.3048	Meters/min.
Feet/min.	0.01136	Miles/hr.
Foot-pounds	0.001286	Btu
Gallons	3785	Cu. centimeters
Gallons	0.1337	Cubic feet
Gallons	231	Cubic inches
Gallons	128	Fluid ounces
Gallons	3.785	Liters
Gallons water	8.35	Lbs. water @ 60°F
Horse-power	42.44	Btu/min.
Horse-power	33.000	Foot-lbs./min.
Horse-power	550	Foot-lbs./sec.
Horse-power	0.7457	Kilowatts
Horse-power	745.7	Watts
Horse-power (boiler)	33.479	Btu/hr.
Horse-power (boiler)	9.803	Kilowatts
Horse -power-hours	2547	Btu
Horse-power-hours	0.7457	Kilowatt-hours
Inches	2.540	Centimeters
Inches	25.4	Millimeters
Inches	0.0254	Meters
Inches	0.0833	Foot
	0.03342	Atmospheres
Inches of mercury	1.133	Feet of water
Inches of mercury	13.57	Inches of water
Inches of mercury	70.73	Lbs./sq. ft.
Inches of mercury	0.4912	Lbs./sq. inch
Inches of mercury	0.002458	Atmospheres
	0.07355	In. of mercury
Inches of water	0.5781	Ounces/sq. inch
Inches of water	5.202	Lbs./sq. loot
Inches of water	0.03613	Lbs./sq. inch
Inches of water	56.92	Btu/min.
Inches of water	1.341	Horse-power
	1000	Watts
Kilowatts	3415	Btu
Kilowatts		
Kilowatts		
Kilowatt-hours		

MULTIPLY	BY	TO OBTAIN
Liters	0.2642	Gallons
Liters	2.113	Pints (liq.)
Liters	1.057	Quarts (liq.)
Meters	100	Centimeters
Meters	3.281	Feet
Meters	39.37	Inches
Meters	1000	Millimeters
Meters	1.094	Yards
Ounces (fluid)	1.805	Cubic inches
Ounces (fluid)	0.02957	Liters
Ounces/sq. inch	0.0625	Lbs./sq. inch
Ounces/sq. inch	1.73	Inches of water
Pints	0.4732	Liter
Pounds (avoir.)	16	Ounces
Pounds of water	0.01602	Cubic feet
Pounds of water	27.68	Cubic inches
Pounds of water	0.1198	Gallons
Pounds/sq. foot	0.01602	Feet of water
Pounds/sq. toot	0.006945	Pounds/sq. inch
Pounds/sq. inch	0.06804	Atmospheres feet of wa-
Pounds/sq. inch	2.307	ter
Pounds/sq. inch	2.036	In. of mercury
Pounds/sq. inch	27.68	Inches of water
Temp. (°C.) + 273	1	Abs. temp. (°C.)
Temp.(°C.) + 17.78	1.8	Temp. (°F.)
Temp. (°F.) + 460	1	Abs. Temp. (°F.)
Temp. (°F) –32	5/9	Temp.(°C.)
Therm	100,000	Btu
Tons (long)	2240	Pounds
Ton, Refrigeration	12,000	Btu/hr.
Tons (short)	2000	Pounds
Watts	3.415	Btu
Watts	0.05692	Btu/min.
Watts	44.26	Foot-pounds/min.
Walls	0.7376	Foot. pounds/sec.
Walls	0.001341	Horse-power
Watts	0.001	Kilowatts
Watt-hours	3.415	Btu/hr.
Watt-hours	2655	Foot-pounds
Watt-hours	0.001341	Horse-power hrs.
Watt-hours	0.001	Kilowatt-hours

FUEL HEATING VALUES

Fuel	*Heating Value*	
Coal		
anthracite	13,900	Btu/lb
bituminous	14,000	Btu/lb
sub-bituminous	12,600	Btu/lb
lignite	11,000	Btu/lb
Heavy Fuel Oils and Middle Distillates		
kerosene	134,000	Btu/gallon
No. 2 burner fuel oil	140,000	Btu/gallon
No. 4 heavy fuel oil	144,000	Btu/gallon
No. 5 heavy fuel oil	150,000	Btu/gallon
No. 6 heavy fuel oil, 2.7 76 sulfur	152,000	Btu/gallon
No. 6 heavy fuel oil, 0.3% sulfur	143,800	Btu/gallon
Gas		
natural	1 000	Btu/cu ft
liquefied butane	103,300	Btu/gallon
liquefied propane	91.500	Btu/gallon

STEAM FORMULAS

PROPERTIES OF SATURATED STEAM

vacuum. Inches of Mercury	Boiling Point or Steam Temperature, Deg. °F.	Specific Volume (V), cu.ft/lb.	$\sqrt{V}$	Maximum Allowable Pressure Drop, pal.(For valve sizing)	Heat of the Liquid. Btu	Latent Heat of Evap., Btu	Total Heat of Steam, Btu
29	76.6	706.00	26.57	0.28	44.7	1048.06	1093.8
26	133.2	145.00	12.04	1.2	101.1	1017.0	1118.1
20	161.2	75.20	8.672	2.4	129.1	1001.0	1180.1
15	178.9	51.30	7.162	3.7	146.8	990.6	1137.4
14	181.8	48.30	6.950	3.9	149.7	988.9	1138.6
12	187.2	43.27	6.576	4.4	155.1	985.6	1140.7
10	192.2	39.16	6.257	4.9	160.1	982.6	1142.7
8	196.7	35.81	5.984	5.4	164.7	980.0	1144.7
6	201.0	32.99	6.744	5.9	168.9	977.2	1146.1
4	204.8	30.62	5.533	6.4	172.8	974.8	1147.6
2	208.5	28.58	5.345	6.9	176.5	972.5	1149.0
Gage Pressure, Pali							
0	212.0	26.79	5.175	7.4	180.0	970.4	1150.4
1	215.3	25.20	5.020	7.8	183.3	968.2	1151.5
2	218.5	23.78	4.876	8.4	186.6	966.2	1152.8
3	221.5	22.57	4.750	8.8	189.6	964.3	1153.9
4	224.4	21.40	4.626	9.4	192.5	962.3	1154.9
5	227.1	20.41	4.518	9.8	195.3	960.6	1155.9
6	229.8	19.45	4.410	10.4	198.0	958.8	1156.8
7	232.3	18.64	4.317	10.8	200.5	957.2	1157.7
8	284.8	17.85	4.225	11.4	208.0	955.5	1158.5
9	237.1	17.16	4.142	11.8	205.4	954.0	1159.4
10	239.4	16.49	4.061	12.4	207.7	952.5	1160.2
11	241.6	15.90	3.987	12.8	209.9	951.1	1161.0
22	243.7	15.35	3.918	13.4	212.1	949.7	1161.8
15	249.8	13.87	3.724	14.8	214.2	948.3	1162.5
20	268.8	12.00	3.464	17.4	227.4	939.5	1166.9
25	266.8	10.57	3.250	19.8	235.6	934.0	1169.6
30	274.0	9.463	3.076	22.4	243.0	928.9	1171.9
40	286.7	7.826	2.797	27.4	255.9	919.9	1775.8
50	297.7	6.682	2.585	32.4	267.1	911.9	1179.0
60	307.3	5.836	2.415	37.4	277.1	904.7	1181.8
70	316.0	5.182	2.276	42.4	286.1	898.0	1184.1
80	323.9	4.662	2.159	47.4	294.3	891.9	1186.2
90	331.2	4.289	2.059	52.4	301.9	886.1	1188.0
100	837.9	3.888	1.972	57.4	308.9	880.7	1189.6
120	350.0	3.337	1.826	67.4	321.7	870.7	1192.4
140	360.9	2.928	1.710	77.4	333.1	861.5	1194.6
160	370.6	2.602	1.613	87.4	343.4	853.0	1196.4
180	379.6	2.345	1.531	97.4	353.0	845.0	1198.0
200	387.8	2.134	1.461	107.4	361.8	837.6	1199.3
250	406.0	1.742	1.320	132.4	381.5	820.2	1201.7
300	421.8	1.472	1.213	157.4	398.6	804.6	1203.2
350	435.6	1.272	1.128	182.4	414.1	790.1	1204.2
400	448.1	1.120	1.068	207.4	428.0	776.6	1204.6
450	459.5	0.998	0.999	232.4	440.9	763.7	1204.6
500	470.0	0.900	0.948	257.4	452.9	751.4	1204.3
550	479.7	0.818	0.904	282.4	464.0	739.7	1203.7
600	488.8	0.749	0.865	307.4	474.6	728.4	1203.0
650	497.3	0.690	0.831	332.4	484.7	717.3	1202.0
700	505.4	0.639	0.799	357.4	494.3	706.5	1200.8
800	520.3	0.554	0.745	407.4	512.3	685.9	1198.2
900	533.9	0.488	0.699	457.4	529.0	666.0	1195.0
1000	546.3	0.435	0.659	507.4	544.2	647.2	1191.4

ELECTRICAL FORMULAS

Three Phase, Alternating Current Motors

$$\text{kW Actual (Motor Input)} = \frac{1.73 \times I \times E \times PF}{1000}$$

$$\text{BHP (Motor Output)} = \frac{1.73 \times I \times E \pm effM \times PF}{746}$$

$$\text{kWh (Motor Input)} = \frac{BHP}{effM}$$

If AMPS are unknown

$$\text{Amps (hp known)} = \frac{BHP \times 746}{1.73 \times E \times effM \times PF}$$

$$\text{Amps (kWh known)} = \frac{kWh \times 1000}{1.73 \times E \times PF}$$

$$\text{Amps (kVa known)} = \frac{kVa \times 1000}{1.73 \times E}$$

Kilo Volt AMPS

$$kVa = \frac{173 \times I \times E}{1000}$$

Motor Electrical Costs For Year

$$\text{Paid kWh Input Per Year} = \frac{173 \times I \times E \times HR}{1000}$$

$$\text{kW Cost Per Year} = \frac{173 \times I \times E \times HR \times \$kWh}{1000}$$

Where

I	=	Current in amps
kWh	=	Kilowatt hours
E	=	Voltage
bhp	=	Break horsepower
PF	=	Power factor
kVa	=	Kilovolt amps
effM	=	Efficiency of motor
FIR	=	Hour per year

REFRIGERATION AND AIR CONDITIONING

TEMPERATURE-PRESSURE CHART

Shaded Figures = vacuum • Solid Figures = Pressure

°F	R-12	R-13	R-22	R-500	R-502	R-717 Ammonia
−100	*27.0*	7.5	*25.0*	*26.4*	*23.3*	*27.4*
−95	*26.4*	10.9	*24.1*	*25.7*	*22.1*	*26.8*
−90	*25.8*	14.2	*23.0*	*24.9*	*20.7*	*26.1*
−85	*25.0*	18.2	*21.7*	*24.0*	*19.0*	*25.3*
−80	*24.1*	22.3	*20.2*	*22.9*	*17.1*	*24.3*
−75	*23.0*	27.1	*18.5*	*21.7*	*15.0*	*23.2*
−70	*21.9*	32.0	*16.6*	*20.3*	*12.6*	*21.9*
−65	*20.5*	37.7	*14.4*	*18.8*	*10.0*	*20.4*
−60	*19.0*	43.5	*12.0*	*17.0*	*7.0*	*18.6*
−55	*17.3*	50.0	*9.2*	*15.0*	*3.6*	*16.6*
−50	*15.4*	57.0	*6.2*	*12.8*	0.0	*14.3*
−45	*13.3*	64.6	*2.7*	*10.4*	2.1	*11.7*
−40	*11.0*	72.7	0.5	*7.6*	4.3	*8.7*
−35	*8.4*	81.5	2.6	*4.6*	6.7	*5.4*
−30	*5.5*	90.9	4.9	*1.2*	9.4	*1.6*
−28	*4.3*	94.9	5.9	0.1	10.5	0.0
−26	*3.0*	96.9	6.9	0.9	11.7	0.8
−24	*1.6*	103.0	7.9	1.6	13.0	1.7
−22	*0.3*	107.3	9.0	2.4	14.2	2.6
−20	0.6	111.7	10.2	3.2	15.5	3.6
−18	1.3	116.2	11.3	4.1	16.9	4.6
−16	2.1	120.8	12.5	5.0	18.3	5.6
−14	2.8	125.7	13.8	5.9	19.7	6.7
−12	3.7	130.5	15.1	6.8	21.2	7.9
−10	4.5	135.4	16.5	7.8	22.8	9.0
−8	5.4	140.5	17.9	8.8	24.4	10.3
−6	6.3	145.7	19.3	9.9	26.0	11.6
−4	7.2	151.1	20.8	11.0	27.7	12.9
−2	8.2	156.5	22.4	12.1	29.4	14.3
0	9.2	162.1	24.0	13.3	31.2	15.7
2	10.2	167.9	25.6	14.5	33.1	17.2
4	11.2	173.7	27.3	15.7	35.0	18.8
6	12.3	179.8	29.1	17.0	37.0	20.4
8	13.5	185.9	30.9	18.4	39.0	22.1
10	14.6	192.1	32.8	19.7	41.1	23.8
12	15.8	196.6	34.7	21.2	43.2	25.6
14	17.1	205.2	36.7	22.6	45.5	27.5
16	18.4	211.9	38.7	24.1	47.7	29.4
18	19.7	218.8	40.9	25.7	50.1	31.4
20	21.0	225.7	43.0	27.3	52.5	33.5
22	22.4	233.0	45.3	28.9	54.9	35.7
24	23.9	240.3	47.6	30.6	57.4	37.9
26	25.4	247.8	49.9	32.4	60.0	40.2
28	26.9	255.5	52.4	34.2	62.7	42.6
30	28.5	263.2	54.9	36.0	65.4	45.0
32	30.1	271.3	57.5	37.9	68.2	47.6
34	31.7	279.5	60.1	39.9	71.1	50.2
36	33.4	287.8	62.8	41.9	74.1	52.9
38	35.2	296.3	65.6	43.9	77.1	55.7
40	37.0	304.9	68.5	46.1	80.2	58.6
45	41.7	327.5	76.0	51.6	88.3	66.3
50	46.7	351.2	84.0	57.6	96.9	74.5
55	52.0	376.1	92.6	63.9	106.0	83.4
60	57.7	402.3	101.6	70.6	115.6	92.9
65	63.8	429.8	111.2	77.8	125.8	103.1
70	70.2	458.7	121.4	85.4	136.6	114.1
75	77.0	489.0	132.2	93.5	148.0	125.8
80	84.2	520.0	143.6	102.0	159.9	138.3
85	91.6	—	155.7	111.0	172.5	151.7
90	99.8	—	168.4	120.6	185.8	165.9
95	106.3	—	181.8	130.6	199.7	181.1
100	117.2	—	195.9	141.2	214.4	197.2
105	126.6	—	210.8	152.4	229.7	214.2
110	136.4	—	226.4	164.1	245.8	232.3
115	146.8	—	242.7	176.5	262.6	251.5
120	157.7	—	759.9	189.4	280.3	271.7
125	169.1	—	277.9	203.0	296.7	293.1
130	181.0	—	296.8	217.2	318.0	315.0
135	193.5	—	316.6	232.1	338.1	335.0
140	206.6	—	337.3	247.7	359.1	365.0
145	220.3	—	358.9	266.1	381.1	390.0
150	234.6	—	381.5	281.1	403.9	420.0
155	249.5	—	405.1	296.9	427.8	450.0
160	265.1	—	429.8	317.4	452.6	490.0

PROPERTIES OF SATURATED STEAM

Centimeters	0.3937	Inches	Horse -power (boiler)	9.803	Kilowatts	Pounds/sq. inch	27.68	Inches of water
Centimeters	0.03280	Feet	**Horse-power-hours**	2547	Btu	**Temp. (°C.) + 273**	1	Abs. temp. (°C.)
Centimeters	0.01	Meters	Horse-power-hours	0.7457	Kilowatt-hours	Temp.(°C.) + 17.78	1.8	Temp. (°F.)
Centimeters	10	Millimeters	**Inches**	2,540	Centimeters	Temp.(°F.) + 460	1	Abs. temp. (°F.)
Centmtrs. of Merc.	0.01316	Atmosphere	Inches	25.4	Millimeters	Temp.(°F) – 32	5/9	Temp.(°C.)
Centimtrs. of merc.	0.4461	Feet of water	Inches	0.0254	Meters	**Therm**	100,000	Btu
Centimtrs. of merc.	136.0	Kgs./sq. meter	Inches	0.0833	Foot	Tons(long)	2240	Pounds
Centimtrs. of merc.	27.85	Lbs./sq. ft.	**Inches of mercury**	0.03342	Atmospheres	**Ton, Refrigeration**	12,000	Btu/hr.
Centimtrs. of merc.	0.1934	Lbs./sq.inch	Inches of mercury	1.133	Feet of water	**Tons (short)**	2000	Pounds
Cubic feet	2.832×10^4	Cubic cms.	Inches of mercury	13.57	Inches of water	**Watts**	3.415	Btu
Cubic feet	1728	Cubic inches	Inches of mercury	70.73	Lbs./sq. ft.	Watts	0.05692	Btu/min.
Cubic feet	0.02832	Cubic meters	Inches of mercury	0.4912	Lbs./sq.inch	Watts	44.26	Foot-pounds/
min.								
Cubic feet	0.03704	Cubic yards	**Inches of water**	0.002458	Atmospheres	Watts	0.7376	Foot-pounds/
sec.								
Cubic feet	7.48052	Gallons U.S.	Inches of water	0.07355	In. of mercury	Watts	0.001341	Horse-power
Cubic feet/minute	472.0	Cubic cms.i'sec.	Inches of water	0.5781	Ounces/sq. inch	Watts	0.001	Kilowatts
Cubic feet/minute	0.1247	Gallons/sec.	Inches of water	5.202	Lbs./sq. toot	**Watt-hours**	3.415	Btu/hr.
Cubic foot water	62.4	Pounds @ 60°F	Inches of water	0.03613	Lbs./sq. inch	Watt-hours	2655	Foot-pounds
Feet	30.48	Centimeters	**Kilowatts**	56.92	Btu/min.	Watt-hours	0.001341	Horse-power
hrs.								
Feet	12	Inches	Kilowatts	1.341	Horse-power	Watt-hours	0.001	Kilowatt-hours
Feet	0.3048	Meters	Kilowatts	1000	Watts			
Feet	1/3	Yards	**Kilowatt-hours**	3415	Btu			

Index

A

Air Balancing
- general procedure, 3-11
- procedure diagram, 12
- proportionate balance, 10
- test reports, 16-19

Air Balance Reports (See test reports, air)
Air Distribution Systems, 107

B

Blank forms, Appendix A, 187

C

Carbon Dioxide
- levels, 89

Computerized Estimating
- features, 153
- piping, 152
- reports, 154
- sheet metal, 152

Cooling Loads, 72-73, 100-101, 221

D

Ductwork
- construction, 108
- estimating, 108
- types, 108

E

Electrical Forms
- electrical consumption, 56-57, 211
- electrical graphs, 60-61
- electrical loads per system, 66-67, 207
- electrical load recap, 68-69, 208

Energy Auditing
- average building energy performance, 48
- diagram, 45
- field surveys, in-depth, 42
- procedures, 41-45
- readings required, 46-47
- sample audit, 49-50
- walk through procedure, 41

Energy Auditing Forms, filled in
- building & system description, 52-54
- chiller test report, 65
- cooling load calculation, 72-73
- cover sheet, 51
- electrical consumption, 56-57
- electrical graphs, 60-61
- electrical loads per system, 66-67
- electrical load recap, 68-69
- energy profiles, 76-77
- equipment, 58-59
- fan test report, 62-63
- heat loss calculation, 70-71
- peak loads per area, 74-75
- pump test report, 64

Energy Auditing Forms, blank (See contents, Appendix A)

Energy Auditing, Residential
- energy consumption history, 78-79
- purpose, 78
- wall, roof, floor, window evaluations, 80-8

Energy Usage
- average US annual, 48

Engineering Calculations
- cooling load calculations, 100-101
- heat loss calculations, 98-99
- peak loads per area, 102-103

Estimating Piping Costs
- computerized piping estimating, 152-155
- fittings and connections, 146
- labor, 147
- pipe and fitting materials, 145
- piping systems, 145
- principles and procedures, 144
- requirements of piping estimator, 144-149
- valves, specialties and equipment, 147

Estimating Piping Forms

piping takeoff sheet, 150-151
Estimating Sheet Metal Costs
computerized sheet metal estimating, 152-155
correction factors, 110
ductwork, 109
equipment, 110
markups, 111
principles and procedures, 106,112-114
procedure diagram, 115
requirements of a sheet metal estimator, 106-114
scope of a sheet metal estimate, 116
wage rates, 111
Estimating Sheet Metal Forms, Filled in
bid check off sheet, 120-122
bid recap sheet, 132-133
bidding record, 134-135
job description and budget costs, 118-119
labor cost per hour, 136-137
markup calculation sheet, 138-139
per lb takeoff sheet, 142-143
per piece duct takeoff sheet, 140-141
quantity takeoff and extension, 126-127
quotations, 124-125
summary sheet, 128-129
telephone quotation, 130-131
Estimating Sheet Metal Forms, blank (See Contents, Appendix A)
Equipment
estimating, 110
fan report, 59
HVAC unit report, 59
pump report, 59
test report data, 11

F
Fans
inspecting, 5
readings, 8-9
test reports, 62-63

H
Heating Loads, 98-99
Humidity
indoor air quality, 90-91
Hydronic Balancing
preparing reports, 28-30
test reports, 32-39
Hydronic Balancing Reports
(See Test Reports, Hydronic)

I
Indoor Air Quality
carbon dioxide, 89
specific causes, 86-87
instruments and testing, 89
formulas, 93-94
humidity, 90-91
outside air, 88-89
particulates, 91-93
problems, 85
testing and instruments, 89-90
Indoor Air Quality Forms filled in, 95-96
Indoor Air Quality Forms blank, 217-219
Instruments
indoor air quality, 89-90

L
Load Calculations
cooling load form, 72-73, 100-101, 221
floor, 84
heat loss form 70-71, 98-99, 220
peak loads per area, 74-75, 102-103, 222
roof and ceiling, 82
wall, 81
windows and doors, 83

M
Motors
checking, 5

O
Outside Air
testing and setting, 88-89

P
Particulate Contaminants, 91-93
aerosols, 93
allergens, 92
fibers, 92
fungal spores, 92

tobacco smoke, 92
Problem Solving and Decision Making
case history, 168
causes, 176
evaluations, 183
information and planning, 182
notes, graphics, 184
objectives, 178
options, 180
procedures, 165
principles, 164
problem, becoming familiar, 173
technical management in, 157
Problem Solving Forms
blank, 238-242
filled in, 157, 177, 179, 181, 182
Proportionate Balancing
air balancing, 10-11
Pumps
reports, 32-33

R

Reports, Procedure for Preparing, 2
Residential Energy Auditing
(See Energy Auditing, Residential)

T

Test Reports, Air, filled in
cover sheet, 13
fan test reports, 16-17
general information, 15
outlet reports, 18-19
Pitot tube traverse, 22-25
system recap, 20-21
Test Reports, Air, blank (See contents, Appendix A)
Test Reports, Hydronic, filled in
chiller, 38-39
procedure, 28-30
pumps, 32-33
water balance, 34-37
Test Reports, Hydronic, blank (See contents, Appendix A)